OXFORD CHEMISTRY PRIMERS

1 S. E. Thomas *Organic synthesis: The roles of boron and silicon*
2 D. T. Davies *Aromatic heterocyclic chemistry*
3 P. R. Jenkins *Organometallic reagents in synthesis*
4 M. Sainsbury *Aromatic chemistry*
5 L. M. Harwood *Polar rearrangements*
6 T. J. Donohoe *Oxidation and reduction in organic synthesis*
7 J. H. Jones *Amino acid and peptide synthesis*
8 C. J. Moody and G. H. Whitham *Reactive intermediates*
9 G. M. Hornby and J. M. Peach *Foundations of organic chemistry*
10 R. Henderson *The mechanisms of reactions at transition metal sites*
11 H. M. Cartwright *Applications of artificial intelligence in chemistry*
12 M. Bochmann *Organometallics 1: Complexes with transition metal–carbon s-bonds*
13 M. Bochmann *Organometallics 2: Complexes with transition metal–carbon p-bonds*
14 C. E. Housecroft *Cluster molecules of the p-block elements*
15 M. J. Winter *Chemical bonding*
16 N. C. Norman *Periodicity and the p-block elements*
17 R. S. Ward *Bifunctional compounds*
18 S. K. Scott *Oscillations, waves, and chaos in chemical kinetics*
19 T. P. Softley *Atomic spectra*
20 J. Mann *Chemical aspects of biosynthesis*
21 B. G. Cox *Modern liquid phase kinetics*
22 A. Harrison *Fractals in chemistry*
23 M. T. Weller *Inorganic materials chemistry*
24 R. P. Wayne *Chemical instrumentation*
25 D. E. Fenton *Biocoordination chemistry*
26 W. G. Richards and P. R. Scott *Energy levels in atoms and molecules*
27 M. J. Winter *d-Block chemistry*
28 D. M. P. Mingos *Essentials of inorganic chemistry 1*
29 G. H. Grant and W. G. Richards *Computational chemistry*
30 S. A. Lee and G. E. Robinson *Process development: Fine chemicals from grams to kilograms*
31 C. L. Willis and M. R. Wills *Organic synthesis*
32 P. J. Hore *Nuclear magnetic resonance*
33 G. H. Whitham *Organosulfur chemistry*
34 A. C. Fisher *Electrode dynamics*
35 G. D. Meakins *Functional groups: Characteristics and interconversions*
36 A. J. Kirby *Stereoelectronic effects*
37 P. A. Cox *Introduction to quantum theory and atomic structure*
38 P. D. Bailey and K. M. Morgan *Organonitrogen chemistry*
39 C. E. Wayne and R. P. Wayne *Photochemistry*
40 C. P. Lawrence, A. Rodger, and R. G. Compton *Foundations of physical chemistry*
41 R. G. Compton and G. H. W. Sanders *Electrode potentials*
42 P. B. Whalley *Two-phase flow and heat transfer*
43 L. M. Harwood and T. D. W. Claridge *Introduction to organic spectroscopy*
44 C. E. Housecroft *Metal–metal bonded carbonyl dimers and clusters*
45 H. Maskill *Mechanisms of organic reactions*
46 P. C. Wilkins and R. G. Wilkins *Inorganic chemistry in biology*
47 J. H. Jones *Core carbonyl chemistry*
48 N. J. B. Green *Quantum mechanics 1: Foundations*
49 I. S. Metcalfe *Chemical reaction engineering: A first course*
50 R. H. S. Winterton *Heat transfer*
51 N. C. Norman *Periodicity and the s- and p-block elements*
52 R. W. Cattrall *Chemical sensors*
53 M. Bowker *The Basis and applications of heterogeneous catalysis*
54 M. C. Grossel *Alicyclic chemistry*
55 J. M. Brown *Molecular spectroscopy*
56 G. J. Price *Thermodynamics of chemical processes*
57 A. G. Howard *Aquatic environmental chemistry*
58 A. O. S. Maczek *Statistical thermodynamics*
59 G. A. Attard and C. J. Barnes *Surfaces*
60 W. Clegg *Crystal structure determination*
61 M. Brouard *Reaction dynamics*
62 A. K. Brisdon *Inorganic spectroscopic methods*
63 G. Proctor *Stereoselectivity in organic synthesis*
64 C. M. A. Brett and A. M. O. Brett *Electroanalysis*
65 N. J. B. Green *Quantum mechanics 2: The toolkit*
66 D. M. P. Mingos *Essentials of inorganic chemistry 2*
67 I. Fleming *Pericyclic reactions*
68 N. S. Lawrence, J. D. Wadhawan & R. G. Compton *Foundations of physical chemistry: Worked examples*
69 J. R. Chipperfield *Non-aqueous solvents*
70 T. J. Mason *Sonochemistry*
71 J. McCleverty *Chemistry of the first-row transition metals*
72 E. C. Constable *Coordination chemistry of macrocyclic compounds*
73 C. E. Housecroft *The heavier d-block metals: Aspects of inorganic and coordination chemistry*
74 P. D. Beer, P. A. Gale and D. K. Smith *Supramolecular chemistry*
76 N. Kaltsoyannis and P. Scott *The f elements*
77 D. S. Sivia and S. G. Rawlings *Foundations of science mathematics*
78 S. Duckett and B. Gilbert *Foundations of spectroscopy*
79 J. H. Atherton and K. J. Carpenter *Process development: Physicochemical concepts*
80 R. de Levie *Acid–base equilibria and titrations*
81 H. Maskill *Structure and reactivity in organic chemistry*
82 D. S. Sivia and S. G. Rawlings *Foundations of science mathematics: Worked problems*
83 J. Iggo *NMR Spectroscopy in inorganic chemistry*
84 P. Biggs *Computers in chemistry*
85 D. J. Walton and P. Lorimer *Polymers*
86 R. J. Davey and J. Garside *From molecules to crystallizers*
87 G. M. Hornby and J. M. Peach *Foundations of organic chemistry: Worked examples*
88 M. J. T. Robinson *Organic stereochemistry*
89 H. R. N. Jones *Radiation heat transfer*
90 J. Saunders *Top drugs—top synthetic routes*
91 M. J. Perkins *Radical chemistry: The fundamentals*
92 P. J. Hore, J. A. Jones, and S. Wimperis *NMR: The toolkit*
93 G. A. D. Ritchie and D. S. Sivia *Foundations of physics for chemists*
94 M. J. Winter and J. Andrew *Foundations of inorganic chemistry*
95 J. Robertson *Protecting group chemistry*
96 R. Whyman *Applied organometallic chemistry and catalysis*
97 J. S. Ogden *Introduction to molecular symmetry*
98 C. M. Dobson, A. J. Pratt, and J. A. Gerrard *Bioorganic chemistry*

Applied Organometallic Chemistry and Catalysis

Robin Whyman

Department of Chemistry, University of Liverpool

Series sponsor: AstraZeneca

This book has been printed digitally and produced in a standard specification in order to ensure its continuing availability

OXFORD
UNIVERSITY PRESS

Great Clarendon Street, Oxford OX2 6DP

Oxford University Press is a department of the University of Oxford.
It furthers the University's objective of excellence in research, scholarship,
and education by publishing world-wide in

Oxford New York

Auckland Bangkok Buenos Aires Cape Town Chennai
Dar es Salaam Delhi Hong Kong Istanbul Karachi Kolkata
Kuala Lumpur Madrid Melbourne Mexico City Mumbai Nairobi
São Paulo Shanghai Taipei Tokyo Toronto

Oxford is a registered trade mark of Oxford University Press
in the UK and in certain other countries

Published in the United States
by Oxford University Press Inc., New York

Reprinted 2003

ISBN 0-19-855917-8

Printed in Great Britain by
Antony Rowe Ltd., Eastbourne

Series Editor's Foreword

Oxford Chemistry Primers are designed to give a concise introduction to all chemistry students by providing the material that would usually form an 8–10 lecture course. As well as providing up-to-date information, this series expresses the explanations and rationales that form the framework of the current understanding of inorganic chemistry.

Robin Whyman has great experience in the field of applied organometallic chemistry and catalysis, having worked in the chemical industry prior to his current appointment at the University of Liverpool. This experience is reflected in this Primer, which provides a remarkable overall perspective of this important field within a single volume. Accordingly, this book will be a great interest to undergraduate, postgraduate, and applied industrial chemists to equal measure. I expect that this Primer will provide an excellent background to Level 3, Level 4, and postgraduate courses in both inorganic, organic, and catalytic chemistry.

John Evans
Department of Chemistry,
University of Southampton

Preface

The phenomenal growth and interest in organometallic chemistry since the discovery of ferrocene in the early 1950s exhibits close parallels with applications in homogeneously catalysed chemical processes for the manufacture of both commodity and fine chemicals. The aim of this Primer is to provide a perspective on these developments over the past half century, with particular emphasis on, first, the influence of organometallic chemistry on our understanding of both homogeneous and heterogeneous catalysis and, second, the principal commercial applications of homogeneous catalysis that have been developed during this period. The content of the Primer has its origins in a Masters degree course presented during my tenure of a visiting professorship at the University of Toronto in the early 1980s. Since then it has been continually developed and updated, and given for several years to Honours chemistry undergraduates at various universities in the UK, most recently at Liverpool. Here, it forms an integral component of the MSc course in surface science and catalysis at the Leverhulme Centre for Innovative Catalysis.

It gives me great pleasure to dedicate this Primer to scientists from the former ICI Corporate Laboratory, Runcorn, in particular Drs J. P. Candlin, W. Hewertson, and D. T. Thompson, who first stimulated my interest in homogeneous catalysis. Finally, I wish to pay tribute to the (considerable!) patience and support of my wife, Pat, during the preparation of the manuscript.

May 2001 — Robin Whyman
Liverpool

Contents

1 Organometallic chemistry and catalysis

1.1 Introduction

Catalysis plays a vital role in the production of fuels, commodity chemicals, fine chemicals, and pharmaceuticals, as well as providing the means for strengthening environmental safeguards all over the world. More than 60% of all chemical products and 90% of chemical processes are based on catalysis. The volume and value of such products began exponential growth in the early 1950s. This coincided with the rapid development of organometallic chemistry and applications in catalytic processes. A whole new technology based on organometallic catalysis appeared and, in some examples such as olefin polymerization, is currently undergoing a renaissance. Nobel prizes have been awarded to Ziegler, Natta, Wilkinson, and Fischer for their contributions to organometallic chemistry and catalysis; most recently (2001), Knowles, Noyori and Sharpless have also been honoured for their pioneering discoveries in asymmetric homogeneous catalysis.

Nobel prizes for Chemistry have been awarded to Ziegler and Natta (1963), Fischer and Wilkinson (1973), Knowles, Noyori and Sharpless (2001), for their pioneering discoveries in organometallic chemistry and catalysis.

Traditionally, heterogeneous (solid) catalysts have been used for the production of large-scale commodity chemicals such as methanol and ammonia, and in the production of high-octane gasoline from petroleum. Homogeneous catalysts, which are soluble in the reaction medium, are assuming ever-increasing significance in the manufacture of both commodity and fine chemicals in high purity. The first application in the former category was their use in the reactions of alkenes with CO and H_2 (hydroformylation) to give aldehydes and alcohols for use in the plasticizer and detergent alcohols businesses. Other major applications of homogeneous catalysts encompass the various production routes to acetic acid and the acetyl family of chemicals, and the oligomerization and polymerization of olefins. Recently, increased emphasis has been directed towards the exploitation of one of the principal advantages associated with homogeneous catalysts, namely their high selectivities, for use in the synthesis of fine chemicals via, in particular, asymmetric catalysis. Several processes for the manufacture of such chemicals have been commercialized and others are under active development.

As indicated in the Preface, the aims of the primer are twofold:

- To provide a perspective on the influence of organometallic chemistry on the development and understanding of both homogeneous and heterogeneous catalysis.
- To provide an account of the major commercial applications of homogeneous catalysis, including those relating to the production of both commodity and fine chemicals.

The first is addressed by a comparison between homogeneous and heterogeneous catalysis (this chapter), and a discussion of the impact of mechanistic aspects derived from organometallic chemistry on both forms of catalysis (Chapter **2**). The interfacial area between the two, for example the role of metal clusters and the concept of surface organometallic chemistry, is included.

The second objective is addressed by a discussion of the development of hydroformylation and related processes (Chapter **3**), the production of acetic acid and the acetyl chemicals (Chapter **4**), the role of buta-1,3-diene hydrocyanation in the production of intermediates in the manufacture of Nylon (Chapter **5**), and the oligomerization, polymerization, co-polymerization and metathesis of olefins (Chapter **6**). A number of these are very well established processes, but some of the most recent developments, for example Ir-catalysed methanol carbonylation (BP Chemicals 'Cativa' technology, 1996) and Pd-catalysed polyketone synthesis from ethylene/CO ('Carilon', Shell, 1996), are included. The increasing number of applications of, in particular, asymmetric homogeneous catalysts in facilitating key steps in the production of fine chemicals, together with factors which have influenced these developments, are illustrated with reference to the manufacture of vitamin A, L-dopa, L-menthol, disparlure, glycidol, ibuprofen, naproxen, and cilastatin (Chapter **7**).

Chapters **3–6** of the primer are sub-divided into process routes towards families of product types rather than specific single chemical reactions. In this way, features such as developmental changes in the processes used for the production of particular chemicals during the past 50–100 years can be highlighted, in addition to the fundamental chemistry associated with the use of homogeneous catalysts. These changes provide an important commentary on general underlying trends in the chemical industry, in particular towards the use of safer, more energy-efficient processes, cheaper feedstocks and with increased concern for environmental issues. In addition to purely chemical aspects, some discussion of factors such as economics, politics and geography, which frequently assume equal, if not greater, importance than chemistry to the processing of industrial chemicals, is included where appropriate.

An important emerging area of applied organometallic chemistry, which is considered to be outside the scope of this primer, relates to applications of organometallic or, more correctly, metallo-organic compounds in the electronics industry. For example, such compounds are finding use as precursors in MOCVD processes for the production of thin films of III–V and II–VI semiconductor materials such as GaAs, InP, and HgCdTe.

1.2 Homogeneous and heterogeneous catalysis

In approaching the subject of applied homogeneous catalysis it is appropriate to begin by comparing and contrasting in general terms the typical properties associated with both homogeneous and heterogeneous catalysts (Table 1.1). The thermodynamic principles of activation are of course the same in both cases but the physical and geometric factors are rather different.

Table 1.1 Homogeneous vs. heterogeneous catalysis

	Homogeneous	*Heterogeneous*
Form	soluble metal complexes, usually mononuclear	metals, usually supported, or metal oxides
Phase	liquid	gas/solid
Temperature	low (<250°C)	high (250–500°C)
Activity	moderate	high
Selectivity	high	low
Diffusion	facile	can be very important
Heat transfer	facile	can be problematic
Product separation	generally problematic	facile
Catalyst recycle	expensive	simple
Reaction mechanisms	reasonably well understood	poorly understood

Heterogeneous catalysts often comprise metals or metal oxides, of which the former may be dispersed on the surfaces of inorganic oxide supports to increase the effective surface area per unit weight of metal component. They are normally used at relatively high temperatures by passing reactants in the vapour phase over the solid catalyst. By contrast, homogeneous catalysts are usually used in the liquid phase at considerably lower temperatures (< 250°C). Such reactions are frequently carried out under pressure to both increase the effective concentration of gaseous reactant(s) in solution, and hence (usually) the reaction rate, and to maintain volatile reactants, e.g. buta-1,3-diene, in the liquid phase under process conditions, with a consequent simplification of chemical engineering requirements.

Activity and selectivity

An inverse relationship between catalytic activity and reaction selectivity often applies.

An inverse relationship between catalytic activity and reaction selectivity is often apparent in both forms of catalysis. Thus a major advantage of homogeneous catalysis is associated with the high selectivities that can be achieved, presumably because the active catalysts are restricted to essentially one type of coordination site. However, this selectivity advantage is frequently achieved only at the expense of reaction rate.

Ni-catalysed olefin dimerization provides an example of high catalytic activity *combined* with high selectivity.

A notable exception to this generalization derives from the elegant work of Wilke *et al.* where *both* high catalytic activities *and* high selectivities were observed in the dimerization of propylene catalysed by phosphine-substituted π-allyl nickel halides after activation with Lewis acids such as $EtAlCl_2$. This reaction was first reported in 1963 and very high catalytic activities of *ca.* 3–6 kg dimer product g^{-1} Ni h^{-1} at −20°C were recorded. Moreover, the dimerization could be controlled within wide limits by the addition of phosphines having different electronic and steric properties (see Chapter **2**), so that dimers of various structures could be prepared at will.

The importance of chemical engineering considerations in addition to elegant chemistry cannot be too highly emphasised.

Subsequently, after careful attention to reaction conditions, including factors such as rates of stirring, cooling and gas feed, it has been established that the intrinsic activity is in fact about a factor of 10^3 *higher*, e.g. 6000–7000 kg product g^{-1} Ni h^{-1}, corresponding to catalytic activities, at room temperature, such as those displayed by enzymes.

Homogeneous catalysts can show activities comparable to those displayed by enzymes.

Developments such as these have given confidence to the view that homogeneous systems hold exciting prospects for the high-technology catalysts of the future.

Product separation

The principal disadvantage with the practical application of homogeneous catalysts is the problem of facile separation of the product from catalyst and reactant on a continuous basis, without incurring irreversible decomposition of the catalyst, and the requirement for expensive catalyst regeneration steps. In some situations, for example, if the product is of low molecular weight and is volatile, e.g. acetaldehyde obtained via Pd-catalysed ethylene oxidation, separation by distillation is facile. In other cases separation problems have been the main hindrance to commercialization of elegant synthetic chemical methodologies. Some creative solutions to this problem have been devised and these are highlighted at appropriate points in the text.

The main problem associated with homogeneous catalysts is separation of the catalyst from reactants and products.

Reaction mechanisms

Mechanistically, homogeneously catalysed reactions are better understood than their heterogeneous counterparts. This is partly because they are kinetically simpler, as a consequence of their higher reaction selectivities, but also because they are more amenable to studies by molecular characterization techniques, for example vibrational and NMR spectroscopies, directly under working conditions. With heterogeneous catalysts, there are obvious difficulties associated with the assessment of intimate mechanisms of adsorption and reaction on a surface, especially under catalytic reaction conditions.

Homogeneous catalysts are readily amenable to kinetic and mechanistic studies, in particular using in situ spectroscopic methods.

Industrial preference for heterogeneous catalysts

Heterogeneous catalysts are the materials of choice for industrial processes, principally on the grounds of higher activity, but also in terms of chemical engineering requirements, such as the ease of separation of products from catalyst and reactants, and robustness, i.e. the ability of the catalyst to withstand vigorous working conditions. The latter property minimizes the possibility of catalyst deactivation or loss by, for example, catalyst plating out from solution onto reactor walls, due to instability caused by local temperature variations or deficiencies in concentrations of gases in solution.

Heterogeneous catalysts are preferred in industrial practice on the grounds of activity, robustness and ease of separation.

Advantages of homogeneous catalysts

Homogeneous catalysts do, however, offer several advantages in addition to the high selectivities referred to above. First, the catalyst is uniform, usually having only one type of reaction site which is not subject to physical surface effects, and may therefore be more reproducible. Second, the catalyst may be readily modified in a controlled manner by variation of parameters such as:

Homogeneous catalysts are uniform, reproducible, and easily modified in a controlled manner.

- metal (type, concentration and oxidation state),
- added ligands (type, ligand:metal ratio, steric and electronic effects),
- added cations and anions, and
- solvent.

With homogeneous catalysts it is frequently possible to characterize individual reaction steps and to isolate key intermediates.

Third, the catalyst is potentially cheaper since, in principle, all metal atoms can be used rather than an unknown, but usually rather small, fraction in heterogeneous catalysis – an important factor if precious metals are involved on a large scale. Finally, kinetic and mechanistic studies are easier with homogeneous catalysts and it is sometimes possible to isolate key intermediates.

Clearly, the areas of homogeneous and heterogeneous catalysis are closely interlinked and in some instances the initial development of a homogeneous catalysed process has later been superseded by a heterogeneous equivalent, as in the oxidation of ethylene to vinyl acetate. By contrast, the hydroformylation reaction was originally believed to be a heterogeneously catalysed process! The majority of commercial processes comprise several stages which are operated in an integrated manner, and a variety of catalyst types may be required. For example, in the three-stage Shell Higher Olefins Process (SHOP), the key step, the oligomerization of ethylene to linear α-olefins, requires a highly selective homogeneous Ni catalyst, whereas the subsequent isomerization and metathesis stages use heterogeneous catalysts.

Finally, it is important to recognize that some chemical transformations are unique to homogeneous catalysts, i.e. no satisfactory heterogeneous analogues are known. These include:

Several chemical transformations can *only* be performed using homogeneous catalysts.

- Pd-catalysed oxidation of ethylene to acetaldehyde (Hoechst Wacker),
- Ni-catalysed hydrocyanation of buta-1,3-diene to adiponitrile (DuPont),
- Rh- and Ru-catalysed reductive coupling of carbon monoxide to ethylene glycol, and
- an increasing number of enantioselective hydrogenation, isomerization and oxidation reactions.

1.3 Commodity chemicals production using homogeneous catalysts

Homogeneous transition metal-catalysed processes that are currently of major significance are summarized in Table 1.2, together with their scale of operation. It will be evident from these entries that oxidation processes such as those involved in the manufacture of intermediates to polyesters occupy a very significant role in homogeneous catalysis, accounting for approximately 9 million tonnes per year (mte y^{-1}) of products. Here, transition metal ions, typically carboxylate salts of Co^{2+} and Mn^{2+}, are used as initiators for the oxidation of *p*-xylene to terephthalic acid (and of cyclohexane to cyclohexanone/cyclohexanol mixtures in Nylon manufacture, see Chapter 5). In such radical-catalysed reactions, the role played by transition metal ions is simply that of radical generation for the initiation of oxidation. As a consequence these are not traditionally included under the 'organometallic' umbrella. However, their commercial significance should not be underestimated. In addition, transition metal salts (and others) are used as catalysts for both transesterification of dimethyl terephthalate and polycondensation of terephthalic acid and ethylene glycol in the manufacture of poly(ethylene terephthalate) (PET).

Oxidation processes for the production of terephthalic acid from *p*-xylene are radical based, and initiated by transition metal ions.

Table 1.2 Production of selected commodity chemicals using homogeneous catalysed processes

Scale of operation worldwide	*mte* y^{-1}
Terephthalic acid and poly(ethylene terephthalate)	9.0
Acetic acid and acetyl chemicals	7.0
Aldehydes and alcohols via hydroformylation	6.0
Adiponitrile via buta-1,3-diene hydrocyanation (DuPont)	1.0
Detergent-range alkenes via Shell Higher Olefins Process (SHOP)	1.0
Total fine chemicals manufacture, including asymmetric catalysis	<1.0
cf. Alkene polymerization (*ca.* 60% production capacity uses Ziegler –Natta type catalysts)	60

The catalysts used for the production of commodity polymers such as polypropylene or polyethylene are often hybrid species, e.g. Ziegler–Natta catalysts (Ti/Al), or those in which an organometallic functional group is bound to a solid support. The distinction between these classes of catalyst as homogeneous or heterogeneous is often rather arbitrary because similar chemical reactions often take place within the coordination sphere of the metal whether it is present in solution or adsorbed on a solid surface. However, in most cases they are not considered to be truly homogeneous catalysts.

Table 1.3 Chronological development of commercial homogeneous catalysed processes

Date	*Transition metal*	*Process*	*Company*
1940s	Co carbonyls	hydroformylation	Ruhrchemie
1950s	Co/Mn acetates	*p*-xylene oxidation	Dynamit Nobel/Hercules, Mid-Century/Amoco
1960	Pd/Cu chlorides	ethylene oxidation	Hoechst–Wacker
1966	Co/phosphine	hydroformylation	Shell
1966	Co/I^-	methanol carbonylation	BASF
1970	Rh/I^-	methanol carbonylation	Monsanto
1971	Ni/phosphite	butadiene hydrocyanation	DuPont
1974	Rh/chiral phosphine	L-dopa	Monsanto
1976	Rh/PPh_3	hydroformylation	Union Carbide/Johnson Matthey/Davy McKee
1977	$Ni/(P^\frown O)$	α-olefins (SHOP)	Shell
1980s	Ti/ROOH/tartrate	(+)-disparlure epoxyalcohols glycidol	May & Baker Upjohn ARCO
1983	$Rh/I^-/[R_4P]I$	methyl acetate carbonylation	Tennessee Eastman
1983	Rh/chiral phosphine	L-menthol	Takasago Int. Corp.
1984	Rh/phosphine/aqueous	hydroformylation	Rhone Poulenc/Ruhrchemie
1989	$Rh/I^-/[R_4N]I$	methanol/methyl acetate co-carbonylation	BP Chemicals
1996	Pd/phosphine	polyketone	Shell
1996	$Ir/I^-/Ru$	methanol carbonylation	BP Chemicals

1.4 Timescales and trends

A chronology of the principal commercial developments in homogeneous catalysis during the past 50–60 years that are referred to in this primer is given in Table 1.3. This enables some generalizations to be drawn concerning the association of particular metals or groups of metals with specific chemical transformations (Table 1.4).

Table 1.4 Catalytic transformations vs. metal type

Transformation	*Metal(s)*
carbonylation and hydroformylation	Co, Rh, Ir, (Pd)
CO/olefin co-polymerization	Pd
ethylene oxidation	Pd
hydrocyanation	Ni
olefin dimerization and oligomerization	Ni (Fe)
olefin polymerization	Ti, Zr, (Ni, Co, Fe)
olefin metathesis	Mo, W

There are, of course, exceptions to every generalization, and it is noteworthy that Rh is probably the most ubiquitous of all transition metals in providing active catalysts for a wide range of transformations, including, in addition to those listed above, olefin dimerization (but *not* polymerization), hydrogenation and isomerization in fine chemicals manufacture (L-dopa, L-menthol respectively).

Rh, one of the most costly platinum group metals, is unique in its activity for a wide range of catalytic transformations.

Perhaps the most surprising absence from the table is that of ruthenium. Apart from possible use in the hydrogenation step in the synthesis of naproxen, Ru-based catalysts have achieved little prominence in applied homogeneous catalysis, although with the development of Ru(BINAP) catalyst systems by Noyori *et al.* (see Chapter 7), a change may be imminent. The Group VI metals Mo and W provide effective homogeneous metathesis catalysts but, in commercial use, heterogeneous versions are preferred. Until the very recent advances utilizing later transition metal complexes containing multidentate N– and N$^\frown$O ligands, high activity for olefin polymerization was believed to be the preserve of early transition metals (see Chapter **6**).

2 Mechanistic organometallic chemistry

2.1 Concepts

A recognition of the vital role played by organometallic chemistry in the understanding of industrial process applications should become readily apparent in later chapters. Developments in organometallic chemistry have frequently occurred in parallel to, and indeed in some cases have been preceded by, industrial applications. The successful commercial application of homogeneous catalysis is based on a good understanding of the underlying organometallic chemistry. In particular, mechanistic concepts derived from organometallic chemistry that are summarized here have not only provided a framework of transformations upon which the construction of catalytic cycles can be based, but have also had important supporting consequences in the substantiation of proposed reaction intermediates present on the surfaces of working heterogeneous catalysts.

18 electron rule

A stable complex (with the electron configuration of the next highest noble gas) is obtained when the sum of the metal *d*-electrons, electrons donated from the ligands, and the overall charge of the complex is 18.

This is an important concept in organotransition metal chemistry, first developed by Tolman in 1972, that may be used in a predictive capacity to derive the most likely stable combinations of metals and ligands in their complexes. It is based on the observation that essentially all well-characterized diamagnetic complexes of the later *d*-block elements contain 16 or 18 metal valence electrons. Tolman proposed two rules for organometallic complexes and their reactions, as follows:

- Diamagnetic organometallic complexes of transition metals may exist in significant concentration at moderate temperatures only if the valence shell of the metal contains 16 or 18 electrons. A significant concentration is defined as one that may be detected spectroscopically or kinetically and may be in the gaseous, liquid or solid state.
- Organometallic reactions, including catalytic ones, proceed by a series of elementary steps involving only intermediates with 16 or 18 valence electrons.

Phosphine ligands

Reference has already been made to the importance of ancillary ligands such as tertiary phosphines in homogeneous catalysis (Section **1.2**). Appropriate selection of functional groups attached to phosphorus can enable systematic variation of electronic and steric properties of these ligands and facilitate fine tuning of catalytic properties.

Tolman introduced the cone angle (θ) and the electronic parameter (χ) to classify phosphine ligands in terms of their steric demand and coordination ability respectively.

Mechanistic aspects

Some of the concepts that originate from organometallic chemistry, and that are of importance to the mechanistic steps required for catalyst activation, reaction and product formation in homogeneously catalysed reactions, are summarized in Table 2.1.

Table 2.1 Mechanistic concepts

Function	*Concept*
Activation of catalyst	coordinative unsaturation – generation of vacant site to accommodate incoming reactant
	'non-classical' coordination of small molecules, e.g. H_2
	oxidative addition – M–C, M–H bond formation; oxidation state of metal formally increases by 2
Reaction on metal centre	alkyl migration/migratory insertion
	alkylidene, metallacycle formation and rearrangement
	nucleophilic and electrophilic addition and abstraction
Release of product from metal centre	reductive elimination – reverse of oxidative addition; oxidation state of metal formally decreases by 2
	β-elimination – typically elimination of an alkene from a metal alkyl complex containing β-hydrogen atoms

Coordinative unsaturation

The reversible, and usually partial, dissociation of a ligand generates a vacant coordination site into which an incoming reactant can be accommodated (Eqns 2.1–2.3).

$$M^xL_n \Longleftrightarrow M^xL_{n-1} + L \qquad (2.1)$$

Direct experimental evidence for coordinative unsaturation is limited and derived largely from kinetic measurements.

$$\underset{18e}{HCo(CO)_4} \Leftrightarrow \underset{16e}{HCo(CO)_3} + CO \quad (2.2)$$

$$\underset{18e}{Pt(PPh_3)_4} \Leftrightarrow \underset{16e}{Pt(PPh_3)_3} \Leftrightarrow \underset{14e}{Pt(PPh_3)_2} \quad (2.3)$$

Non-classical coordination of small molecules

The use of transition metal complexes containing bulky ligands allows the stabilization and isolation of unusual structures (Eqn 2.4).

$$M(CO)_3L_2 + H_2 \Leftrightarrow M(CO)_3(\eta_2\text{-}H_2)L_2 \quad (2.4)$$

M = Mo,W; L = bulky phosphine ligand: PPr^i_3, PCy_3 (Cy = cyclohexyl)

The weak, and reversible, co-ordination of small molecules such as H_2 can be considered to resemble the phenomenon of physical adsorption in heterogeneous catalysis.

First discovered in 1984, the formation of coordinated molecular dihydrogen complexes, in which the bond between the hydrogen atoms is retained, can be considered to represent the incipient oxidative addition (see below) of a diatomic molecule. Such complexes are often only weakly stabilized and readily regenerate the starting complex by replacing the H_2 source with, for example, N_2. For M = W and L = PPr^i_3, the presence, in solution, of an equilibrium between the dihydrogen complex and the product of oxidative addition of hydrogen, namely $W(CO)_3(H)(H)L_2$, has been established.

Oxidative addition

This generalized reaction (Eqn 2.5) provides the most important route for the formation of metal–carbon and metal–hydrogen bonds in organometallic chemistry.

$$M^xL_n + X\text{–}Y \Leftrightarrow M^{x+2}L_n(X)(Y) \quad (2.5)$$

X = H; Y = H, Cl, $HC{=}CR_2$, $C{\equiv}CR$

X = alkyl, acyl, aroyl; Y = halogen

Typical examples of oxidative addition, with the formation of M–H and M–C bonds, respectively, are shown in Eqns 2.6 and 2.7.

$$Ir^I(CO)Cl(PPh_3)_2 + H_2 \longrightarrow (H)(H)Ir^{III}(CO)Cl(PPh_3)_2 \quad (2.6)$$

$$[Ir^{I}(CO)_2I_2]^- + CH_3I \longrightarrow [CH_3Ir^{III}(CO)_2I_3]^- \quad (2.7)$$

Requirements for oxidative addition include the availability of two additional coordination sites and the ability of the metal to accommodate an increase in oxidation state by two, e.g. Rh(I)—>Rh(III), Pt(0)—>Pt(II), etc.

The oxidative addition reaction has some additional electronic and steric limitations imposed upon it relative to the other steps described in this section since both the oxidation state of the metal and the coordination number in the resulting product are increased by two.

Migratory insertion

The generalized migratory insertion reaction (Eqn 2.8), also termed alkyl migration and carbonyl insertion, depending on the nature of the functional groups L and X, is of wide scope (Table 2.2).

$$X—M^{x}—L \longrightarrow M^{x}—X—L \quad (2.8)$$

Table 2.2 Functional groups which may undergo migratory insertion reactions

M–L bond	*Group inserted X*
M–C	CO, CNR, CR_2, C=C, C≡C, CO_2, O_2, SO_2, $SnCl_2$
M–H	CO_2, CNR, C=C, RNCO

Migratory insertion assumes particular importance in reactions involving carbon monoxide, for example the hydroformylation and carbonylation of olefins and alcohols, respectively. The classic example is the methyl migration mechanism established for the conversion of $CH_3Mn(CO)_5$ into the acyl complex (Eqn 2.9).

Migratory insertion is of particular importance in reactions involving the incorporation of CO into organic molecules.

$$R\text{-}Mn(CO)_5 \Longleftrightarrow [RCOMn(CO)_4] \xrightarrow{*CO} RCOMn(CO)_4(^*CO) \quad (2.9)$$

Formation of metallacyclic intermediates

These are important in the facilitation of hydrocarbon rearrangements, e.g. metathesis, where an interchange of carbon frameworks can occur:

$$L_nM{=}CR_2 + \begin{matrix} C^* \\ | \\ C^* \end{matrix} \Longleftrightarrow L_nM\begin{matrix} \diagup C(R_2) \diagdown \\ \diagdown C^* \diagup \end{matrix}C^* \Longleftrightarrow L_nM{=}C^* + \begin{matrix} CR_2 \\ | \\ C^* \end{matrix}$$

Many examples of discrete metallacyclic complexes that contain a range of ring sizes and metal centres, e.g. mono- and di-metallocyclobutanes, -pentanes, -hexanes, etc., are known.

Nucleophilic and electrophilic addition and abstraction

The activation of an unsaturated ligand, coordinated to a metal centre, towards direct attack of an external reagent with, or without, prior binding of the reagent to the metal may be promoted by either nucleophilic or electrophilic reagents. For example, in the Pd-catalysed oxidation of ethylene, in acetic acid, coordination of ethylene to the Pd(II) centre decreases the electron density on the olefin and renders it more susceptible to intramolecular nucleophilic attack by acetate:

$$\mathrm{Pd}(\eta^2\text{-}CH_2{=}CH_2)(OAc) \longrightarrow \mathrm{Pd}{-}CH_2CH_2OAc$$

Similarly, the coordination of olefins to organometallic entities such as $[(\pi\text{-}C_5H_5)Fe(CO)_2{-}]$ renders them susceptible to attack by a wide range of nucleophiles under mild reaction conditions. Typical nucleophiles are $[CH(CO_2Et)_2]^-$, R^-, X^-, etc. In the majority of cases this activation is principally electronic in character. Nevertheless, steric considerations can be equally, and in some cases are overridingly, important in activation by coordination processes – this is especially true in asymmetric synthesis and catalysis (see Chapter 7).

Reductive elimination

Mechanistically reductive elimination is generally not well understood.

This is essentially the reverse of oxidative addition, and usually the last step in a catalytic cycle, which involves the release of product and regeneration of the catalyst, e.g. hydrogenation and hydroformylation, Eqns 2.10 and 2.11.

$$R(H)M^xL_n \longrightarrow RH + M^{x-2}L_n \qquad (2.10)$$

$$RCO(H)M^xL_n \longrightarrow RCHO + M^{x-2}L_n \qquad (2.11)$$

***β*-elimination**

In this case a transition metal alkyl complex $M{-}CH_2CH_2R$, containing a *β*-hydrogen atom, undergoes a 1,2-hydrogen shift with the formation of a metal hydride and elimination of an alkene (Eqn 2.12).

$$M{-}CH_2CH_2R \longrightarrow M{-}H + CH_2{=}CHR \qquad (2.12)$$

β-Hydride elimination was a significant contributory factor to the difficulties in the isolation of stable transition metal alkyl complexes during the early development of organometallic chemistry. This reaction is of particular significance as a termination step in transition metal catalysed olefin polymerization reactions (see Chapter **6**) and during olefin isomerization.

2.2 Construction of catalytic cycles

Appropriate combinations of some of the concepts outlined above, together with a consideration of Tolman's 18 electron rule, may be used to assemble simple catalytic cycles. As an example (Fig. 2.1), for olefin hydrogenation using Wilkinson's catalyst, $RhCl(PPh_3)_3$, the cycle incorporates, clockwise from the top:

- oxidative addition of H_2 to the coordinatively unsaturated Rh(I) complex $RhCl(PPh_3)_2$,
- coordination of olefin to the resultant Rh(III) dihydride,
- migratory insertion of olefin into one Rh–H bond to generate an alkylrhodium(III) monohydride, and
- reductive elimination, with simultaneous release of the product of hydrogenation and regeneration of the Rh(I) catalyst.

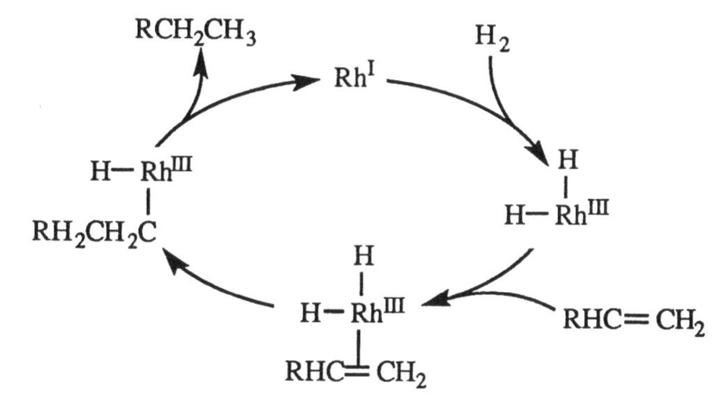

Fig. 2.1 Example of a simple catalytic cycle for olefin hydrogenation, with ancillary ligands removed for clarity.

A catalytic system can thus be seen as a series of reactions connected in such a manner that during one cycle around the loop a substrate is converted into product(s); no *net* change is demanded of the catalyst itself. Catalytic cycles can be constructed either retrospectively, in an attempt to rationalize a known catalytic process, or prospectively, as an aid to catalyst design/discovery.

2.3 Heterogeneous catalyst counterparts

As an extension to the discussion in Section **2.1**, a parallel between most of the concepts summarized in Table 2.1 and the events that are presumed to occur on the surfaces of heterogeneous catalysts can be developed (Table 2.3).

Table 2.3 Parallels between mechanistic 'events' in homogeneous and heterogeneous catalysis

Homogeneous	*Heterogeneous*
	diffusion of reactants to surface
coordinative unsaturation – generation of vacant sites	availability of active surface sites
'non-classical' coordination of small molecules, e.g. H_2	physical adsorption (physisorption)
oxidative addition with chemical bond formation	chemical adsorption (chemisorption)
migratory insertion, metallacycle formation and rearrangement, nucleophilic and electrophilic addition and abstraction	reaction on surface
reductive elimination, β-elimination	desorption of product(s)
	diffusion of product(s) away from surface

Thus, the reversible formation of 'non-classical' dihydrogen complexes, in which the H–H bond remains intact, can be likened to the phenomenon of *physical adsorption*, whereas the 'classical' reaction involving H–H bond cleavage with concomitant oxidative addition to the metal centre can be compared to the process of *chemisorption*. Similarly, the reverse process, reductive elimination, has a parallel in *product desorption* from the surface of a heterogeneous catalyst.

The intimate nature of reactions occurring on the surfaces of heterogeneous catalysts is usually far from clear but the individual reaction types and steps that have been identified from studies of organometallic chemistry and homogeneous catalysis can provide pointers concerning reaction pathways. In addition, the multitude of organometallic complexes containing a myriad of hydrocarbon fragment configurations that have been unambiguously characterized by X-ray crystallography provides convincing supporting evidence for many of the proposed reaction intermediates on the surfaces of heterogeneous catalysts. The apparent links between organometallic chemistry and heterogeneous catalysis can be extended further by consideration of:

- the role of metal clusters in acting as a bridge between conventional homogeneous and heterogeneous catalysis, and
- the development of the concept of surface organometallic chemistry.

2.4 Metal clusters

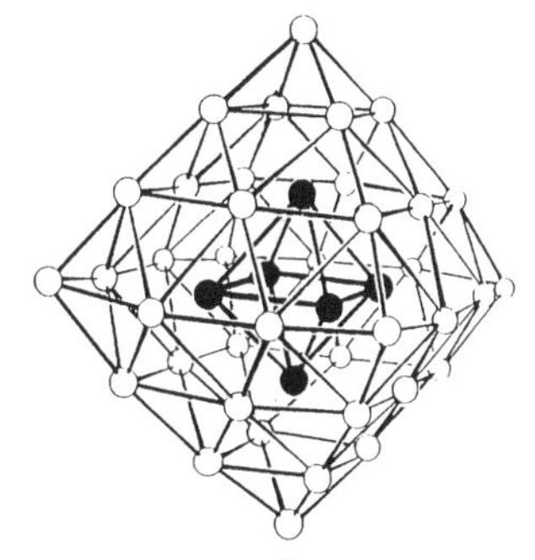

$[Ni_{38}Pt_6(CO)_{48}H]^{5-}$ metal skeleton
o = Ni ● = Pt

A convenient link between the areas of conventional homogeneous and heterogeneous catalysis, particularly in the context of catalysis by metals, has been provided by developments in the area of metal clusters. As described in Section **1.2**, homogeneous catalysts typically contain soluble mononuclear complexes whereas catalysis by supported metals usually involves metals in the bulk solid state, i.e. comprising thousands of atoms. For example, in a typical transition metal catalyst containing metal particles of size 5 nm, each individual particle may contain *ca.* 3000 atoms. A bridge between these two extreme situations has been provided by the efforts of organometallic chemists in the preparation and characterization of a vast range of both homo- and hetero-metallic cluster compounds containing a variety of coordinated ligands including carbon monoxide, hydrocarbyl fragments, etc.

Among the larger transition metal cluster compounds to have been characterized by X-ray crystallography are $[Ni_{38}Pt_6(CO)_{48}H]^{5-}$, in which the Ni atoms are arranged around a central Pt core in the cubic close-packed structure that is typical of many bulk metals, and, very recently (Nov 2000), the 'three-shell' cluster $[Pd_{145}(CO)_x(PEt_3)_{30}]$. By using such well-characterized cluster compounds as catalyst precursors it is possible to envisage a scenario (Table 2.4) in which clusters in various guises act as a link between the conventional disciplines of homogeneous and heterogeneous catalysis.

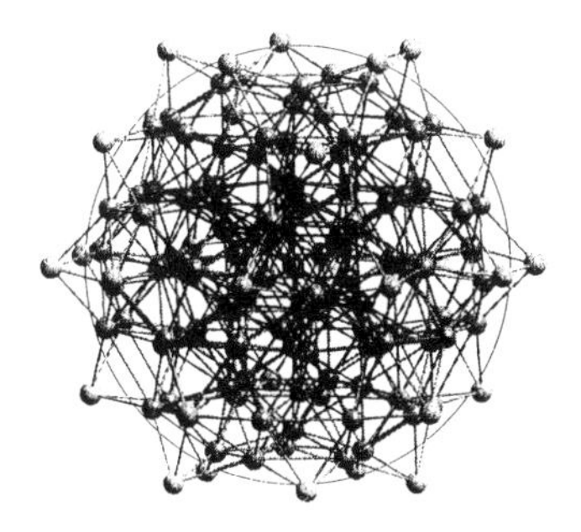

$[Pd_{145}(CO)_x(PEt_3)_{30}]$ three-shell metal cluster skeleton

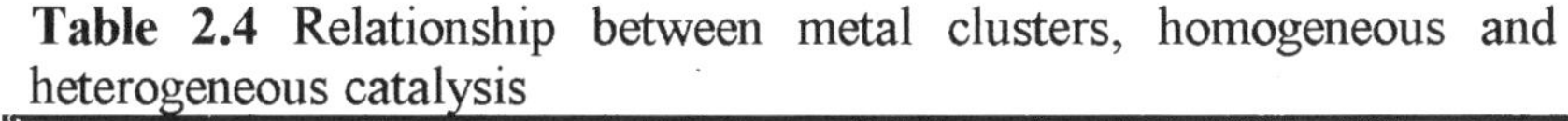

Table 2.4 Relationship between metal clusters, homogeneous and heterogeneous catalysis

Conventional homogeneous catalysis	ML_n */solvent*
homogeneous cluster catalysis	M_xL_n/solvent
heterogeneous supported clusters	$M_y L_n$/support
supported metal nanoparticles	M_y/support
conventional heterogeneous catalysis	M_∞/support

It must, however, be emphasized that in spite of extensive efforts and many exhortations to the contrary, very few, if any, unequivocal examples of true homogeneous catalysis by molecular metal clusters are known. Fragmentation of metal clusters to mononuclear species – sometimes reversibly – is frequently observed under catalytic reaction conditions, particularly if CO is one of the reaction components. Rather more promising results have been obtained from the study of the catalytic behaviour of supported metal clusters, which is closely related to the concept of surface organometallic chemistry.

2.5 Surface organometallic chemistry

Surface organometallic chemistry is a term which has been coined relatively recently and which also represents a bridge between classical homogeneous and heterogeneous catalysis. By using well-characterized molecular organometallic complexes such as metal alkyls, and reacting them in a defined way with surface species (e.g. ≡Si–OH groups), molecularly dispersed, but presumed immobile, catalyst species can be produced in a well-defined chemical environment. The silica-supported alkylzirconium(IV) surface species that efficiently hydrogenates olefins is an example:

```
                                                              CH2C(CH3)3
                                                              |
                                                              Zr
                                                            / | \
 OH    OH    OH       Zr[CH2C(CH3)3]4                      O  O  O
 |     |     |        ——————>                              |  |  |
–O–Si–O–Si–O–Si–O–    – 3C(CH3)4                      –O–Si–O–Si–O–Si–O–
//////////////////////                                //////////////////////
```

A commercial application of the concept of surface organometallic chemistry, some 20 years in advance of the terminology(!), is the anchoring of $(\pi\text{-}C_5H_5)_2Cr$ to the surface hydroxyl groups of silica in the formation of precursors for a range of ethylene polymerization catalysts (see Section **6.4**).

Application of the concept of surface organometallic chemistry may thus facilitate the formation of well-defined surface species (molecular analogues) on high surface area solids such as silica, alumina and magnesia. These offer the prospect of new catalysts in their own right, including, for example, supported analogues of known molecular (homogeneous) catalysts, metal centres stabilized in low or high oxidation states, and multifunctional tailor-made sites such as metal–promoter combinations. Surface-bound organometallic complexes can also serve as precursors of small metal particles (Section **2.4**), including compositionally uniform bimetallic particles.

Nevertheless it is important to recognize that in this interfacial area between homogeneous and heterogeneous catalysts, attempts to 'heterogenize' homogeneous processes have in reality met with very limited success in commercial practice, principally as a consequence of slow (down to ppb level) leaching of catalytic metals from supports, under the sometimes extreme reaction conditions required in working processes.

It is anticipated that the information presented in these two introductory chapters will provide adequate background material from which to provide a perspective on the details of the commercial homogeneously catalysed processes described in Chapters **3–7**.

3 Hydroformylation and related reactions

3.1 Introduction

Hydroformylation, or the OXO (oxonation) reaction, is the name used to describe the addition of CO and H_2 to the C=C bond of an olefin, or other unsaturated compound, with the formation of aldehydes containing one more carbon atom than the olefin (Eqn 3.1).

$$RCH{=}CH_2 + CO + H_2 \longrightarrow RCH_2CH_2CHO + RCH(CH_3)CHO \quad (3.1)$$

With the exception of the case in which R = H an isomeric mixture of straight- and branched-chain aldehydes is produced.

The reaction was discovered in 1938 by Roelen at Ruhrchemie while investigating the recycling of olefins in the heterogeneously catalysed Fischer-Tropsch synthesis. Since that time, the reaction has been developed to form the basis of an industry which, worldwide, accounts for *ca.* 6 million tonnes per year of products. The aldehydes which are produced in the reaction are not normally used as such, but are reduced to alcohols using heterogeneous catalysts for three principal applications, namely solvents, plasticizers, and detergents (Table 3.1).

Table 3.1 Most common raw materials and end-uses of hydroformylation products

Raw material	*Hydroformylation product*	*Alcohol*	*End-use*
Propylene	*iso*-butyraldehyde	*iso*-butanol	solvents
	n-butyraldehyde	*n*-butanol	lacquer industry
	n-butyraldehyde	2-ethylhexanol	plasticizers
Hept-1-ene	octanaldehydes	octanols	plasticizers
C_7–C_{10} olefins	C_8–C_{11} aldehydes	C_8–C_{11} alcohols	plasticizers
C_{12}–C_{14} olefins	C_{13}–C_{15} aldehydes	C_{13}–C_{15} alcohols	synthetic detergents

Production of the plasticizer alcohol 2-ethylhexanol from *n*-butyraldehyde is a particularly important application of the hydroformylation of propylene.

The hydroformylation of propylene merits particular mention because this comprises the principal route to *n*-butyraldehyde, an important intermediate in the manufacture of 2-ethylhexanol. Base-catalysed aldol condensation, followed by dehydration and hydrogenation yields 2-ethylhexanol, which, after conversion to its phthalate ester, has particularly desirable properties as a plasticizer for PVC.

Other important applications include the synthesis of fatty alcohols from terminal olefins such as hept-1-ene. Linear alcohols prepared by hydroformylation of, for example, mixtures of C_{12} olefins obtained by controlled oligomerization of propylene, are used in the synthesis of biodegradable detergents.

Hydroformylation is applicable to a wide range of unsaturated substrates.

Although the bulk of production via hydroformylation involves relatively simple olefins, the reaction may also be applied to a great variety of substituted olefins, both simple and complex, including unsaturated oils, fats and polymers, as well as unsaturated cyclic compounds such as terpenes, pyrans, carbohydrates and steroids. The reaction is utilized commercially in the synthesis of fine chemicals such as vitamin A (see Section **7.2**).

3.2 Hydroformylation processes

Hydroformylation was the first homogeneously catalysed process to find industrial use, although this was neither realized nor established until a number of years after its discovery.

The basic hydroformylation reaction is exothermic, with a heat of reaction for propylene of 125 kJ mol^{-1} and 115–145 kJ mol^{-1} for other olefins, depending upon olefin structure and molecular weight, and is thermodynamically favourable at ambient pressures and low temperatures. However, it required over 35 years after the discovery of the reaction before these mild conditions were realized in practice. The reaction proceeds only in the presence of metal carbonyl catalysts and is the most important industrial synthesis to use these. Although heterogeneous catalysts [Co(30%)/SiO_2(66%)/ThO_2(2%)MgO(2%)] were used initially, the reaction is in fact homogeneously catalysed.

In spite of many attempts, success at the development of commercially viable heterogeneous hydroformylation catalysts has been elusive. This is because of slow leaching of catalyst from the support under process conditions, as a consequence of the equilibrium between supported metal and CO being displaced strongly in favour of the free metal carbonyl.

There are essentially four variations of olefin hydroformylation processes that have been commercialized, namely:

- the Co-catalysed process
- the phosphine-modified Co process
- the phosphine-modified Rh process, and
- a biphasic version of the latter using water-soluble phosphines.

The reaction conditions and essential process parameters for each of these are summarized in Table 3.2.

The discovery and development of these largely complementary processes, each of which has its own niche, with inherent advantages and disadvantages, exhibit striking parallels with progress in organometallic chemistry.

It should be emphasized that the first-generation cobalt carbonyl catalysts are still used for the major proportion of C_n (n>4) aldehyde manufacture.

Table 3.2 Commercial hydroformylation processes – reaction conditions and process parameters

Process	*Co*	*Co/P*	*Rh/P*	*Rh/P(biphasic)*
Catalyst precursor	$Co_2(CO)_8$ or Co salts	$Co_2(CO)_8/PR_3$ R = Bu^n, other	$HRh(CO)(PPh_3)_3$	$HRh(CO)(PR_3)_3$ R = m-$C_6H_4SO_3Na$
Phosphine: metal ratio	–	2:1	50:1 —> 100:1	50:1 —> 100:1
Pressure (bar)	200–300	50–100	15–25	40–60
Temperature (°C)	110–160	160–200	80–120	110–130
Catalyst concentration (% metal/olefin)	0.1–1.0	0.6	0.01–0.05	0.001–1
n/*iso* product ratio	4:1	7:1	8:1 —> 16:1	7:1 —> 19:1
Olefin hydrogenation (%)	< 2	15	5	< 2
High boiling products (%)	5	5	2	< 0.5
Catalyst recovery and recycle	difficult	simpler	simple only for C_3 and C_4 olefins	facile

A $CO:H_2$ molar ratio of *ca.* 1:1 is optimal for aldehyde production.

Several key points that apply to each of the entries of process conditions and parameters in Table 3.2 require amplification. As is apparent from Eqn 3.1, with the exception of R = H, an isomeric mixture of straight- and branched-chain aldehydes, dependent on the extent of *anti*-Markovnikov vs. Markovnikov addition, is produced during hydroformylation. The composition of this mixture, which is commonly defined as the *n*/*iso* ratio, is one of the most important features of a hydroformylation process. As a general rule, maximum selectivities to straight-chain products are preferred.

A key feature of hydroformylation is that high selectivities to straight-chain(*n*) rather than branched-chain (*iso*) aldehydes are generally preferred.

This is particularly important if one of the products is of higher value, as is the case with the hydroformylation of propylene to *n*- and *iso*-butyraldehyde. It is not as important if the mixtures are converted together and are equivalent in the final products, as is the situation with, for example, the conversion of di-*iso*-butylene or tri-propylene to plasticizer alcohols. Another important feature to emerge from Table 3.2 concerns undesirable competing side-reactions such as the hydrogenation of the starting olefins (and aldehydes) to paraffin hydrocarbons (and alcohols) and the formation of high boiling products (termed 'heavy ends') from condensation reactions of the product aldehydes. Aldehyde hydrogenation is not important if the total mixture of isomers is subsequently hydrogenated to alcohols but if aldehydes are the primary requirement, as with *n*-butyraldehyde for subsequent conversion into 2-ethylhexanol, then minimal alcohol formation is required in the primary hydroformylation step.

Side-reactions such as olefin hydrogenation and aldehyde condensation reduce the overall efficiency of hydroformylation.

The Co-catalysed process

In the classical Co-catalysed process many forms of cobalt may be used as catalyst precursor, e.g. the carbonyl, metal, hydroxide, oxide, carbonate, sulphate, fatty acid salt, and Raney Co, all of which are believed to be converted into common Co carbonyls and hydridocarbonyls under operating conditions. The reaction conditions are rather severe, particularly in respect of the high pressure requirement which in turn has, until relatively recently, considerably inhibited accurate kinetic and mechanistic studies. Adequate *n/iso* ratios are typical and the extent of by-product formation is limited.

The Co-catalysed process provides adequate *n/iso* ratios and some flexibility to fine-tune by variation of reaction conditions.

Whereas the *n/iso* product ratio appears to be relatively independent of catalyst concentration and solvent, it does exhibit some dependence on temperature and partial pressures of CO, the use of lower temperatures and higher P_{CO} favouring the formation of the straight-chain isomer, although with a concomitant decrease in overall conversion. These effects are probably caused by shifts in the equilibria between straight- and branched-chain alkyl and acyl Co carbonyls (see Section **3.3**). By making use of these variables it was possible to exercise some control over the *n/iso* ratio, within certain limits, and to operate the process successfully for *ca.* 20 years without the necessity for technical innovation and improvement. However, one significant problem did prove difficult to overcome, and this is a direct consequence of the thermal instability of Co carbonyls and their tendency to deposit from solution onto reactor walls as either metal or metal oxides, thus exemplifying one of the principal problems associated with the commercial operation of homogeneous catalysts (Section **1.2**). In addition to loss of catalytic activity, this also resulted in significant problems with catalyst re-use and necessitated the inclusion of a capital intensive catalyst recycle stage in the process. Overall, therefore, a number of incentives to improve the original Co-based process became apparent, namely:

- operation at lower pressures, thus reducing the high inherent capital costs associated with compressors,
- improvement of the *n/iso* aldehyde product ratio,
- enhancement of catalyst stability under operating conditions,
- improvement of catalyst recovery, and
- reduction of competing side-reactions to a minimum.

The problem of thermal instability of Co carbonyls and deposition on reactor walls over time, leading to loss of catalytic activity and costly regeneration, provided a major incentive to develop alternative process options.

As a first step in this direction a very obvious possibility was to consider the catalytic activity of metal carbonyls other than cobalt. The outcome of such a comparative study (Table 3.3), with activities normalized against a value of 1 for Co, shows that Co and Rh are the metals of choice for hydroformylation.

Table 3.3 Relative activities (in parentheses) of metal carbonyls as hydroformylation catalysts

Mn(10^{-4})	Fe(10^{-6})	Co(1)	Ni($<10^{-6}$)
	Ru(10^{-2})	Rh(10^{3}–10^{4})	Pd
Re	Os	Ir(10^{-2})	Pt

Hydroformylation activities of metal carbonyls: Rh >> Co >> Ir, Ru.

Nevertheless, in spite of the catalytic activity advantages associated with Rh, there are two major disadvantages, namely cost and an *n/iso* aldehyde product ratio of only 1:1, compared to 4:1 in the case of Co. Other metal carbonyls are considerably less active and it is at first sight surprising, considering the origins of hydroformylation, that the other traditional Fischer–Tropsch catalysts, Fe and Ni, are so much less active than Co. Thus no commercially viable alternative to Co emerged directly from this comparative study.

Although Rh is by far the most active metal, the *n/iso* product ratio is only 1:1.

In parallel with the development of organometallic chemistry during the late 1950s and early 1960s it became apparent that replacement of some of the carbonyl groups of metal carbonyls by, in particular, tertiary phosphine and arsine ligands, resulted in the formation of complexes of higher thermal stability. This provided a fundamental advance, which has been subsequently applied to many homogeneously catalysed reactions, by providing the capability to fine-tune and tailor-make catalysts by variation of both the electronic and steric properties of the attached ligands (Sections **1.2** and **2.1**). The ready availability of

The recognition of the enhanced thermal stability displayed by phosphine-substituted metal carbonyls, and the ability to fine-tune catalytic activity and selectivity by variation of functional groups on the ligands, led to the development of a second generation of hydroformylation processes.

a wide range of such ligands led to the next generation of hydroformylation catalysts, as exemplified by the development of the Shell phosphine-modified cobalt and the Union Carbide/Johnson Matthey/Davy McKee phosphine-modified rhodium processes, commercialized in 1966 and 1976 respectively.

The phosphine-modified Co process

Characteristics of the phosphine-modified process include lower operating pressures, improved *n/iso* ratios, but lower activities; the presence of the phosphine ligand partially poisons the Co catalyst.

The enhanced hydrogenation capacity afforded by phosphine substitution can be profitably utilized in the direct formation of alcohols.

A comparison between the key features of the classical Co and phosphine-modified Co processes (Table 3.2) clearly demonstrates the attendant advantages and disadvantages. The addition of a tertiary phosphine has provided a catalyst with increased thermal stability which, in turn, facilitates operation at lower pressures and higher temperatures. The relatively high thermal stability of the catalyst also facilitates the separation of products from the catalyst by distillation. In addition, product linearity is significantly enhanced even when the feedstock comprises *internal* linear olefins. However, these advantages are offset by both a considerable reduction in reaction rate (at 180°C the rate is only 20% of that of conventional Co-catalysed processes at 145° C) and the fact that the hydrogenation capacity of the system has been enhanced to such an extent that paraffin formation becomes a significant side-reaction. If alcohols rather than aldehydes are the required products, this can be a very useful process. Indeed, Shell commercialized the process specifically for the conversion of medium chain length C_7–C_{14} terminal olefins into linear alcohols sold under the tradename 'Dobanol'.

When the hydroformylation is carried out in an alkaline reaction medium, propylene can be converted directly into 2-ethylhexanol.

In addition, propylene can be directly converted into 2-ethylhexanol when the hydroformylation is carried out in an alkaline reaction medium, by the addition of KOH or a tertiary alkylamine. Sequential hydroformylation, aldol condensation, dehydration and hydrogenation occur to give 2-ethylhexanol, in yields of up to 85%, in a single reactor.

The phosphine-modified Rh process

The phosphine-modified Rh process is effective under very mild reaction conditions.

A comparison of the essential features of this process with those of the Co predecessors (Table 3.2) exemplifies the continued trend towards milder reaction conditions and increased linearity of products. Indeed, this catalyst is effective at ambient pressure although, as a consequence of chemical engineering requirements, the commercial process is operated at *ca.* 20 bar total pressure. The *n/iso* ratio is dependent upon the amount of added phosphine but can be as high as 16:1, corresponding to a 94% selectivity to linear products. As a consequence of the mild reaction conditions only small amounts of high boiling products are generated, although olefin hydrogenation is a significant side-reaction. The catalyst concentration is low compared to both Co processes but against this has to be set the cost and availability of Rh, together with maintenance of Rh losses to an absolute minimum (<0.01%).

This process, although excellent for the hydroformylation of lower olefins such as propylene, is of extremely limited use for the hydroformylation of higher olefins because of thermal instability of the catalyst at the high temperatures required for removal of the product aldehydes by distillation (Section **1.2**). Thus, at this stage of development the process is less versatile than either of the Co-based processes. Nevertheless the mild reaction conditions under which this catalyst operates make the phosphine-modified rhodium system particularly attractive for the production, on a (relatively) small scale, of high-added-value fine chemicals.

This process is typified by high (and tunable) *n/iso* ratios and mild operating conditions, but for commodity chemicals manufacture is limited to the hydroformylation of propylene.

The biphasic phosphine-modified Rh process

In attempts to redress difficulties in the application of the phosphine-modified Rh process to higher olefins, the most recent development in hydroformylation technology concerns the use of a water-soluble phosphine such as $P(m\text{-}C_6H_4SO_3Na)_3$, obtained by sulphonation of PPh_3, as ligand. This effectively facilitates hydroformylation in the aqueous phase; the product aldehydes are immiscible and can therefore be readily separated. The concept was commercialized by Rhone Poulenc–Ruhrchemie in 1984 and is now operated in four plants with a production capacity of 450,000 te y^{-1}. Reaction conditions are similar to those used in the initial version of the phosphine-modified Rh process (Table 3.2). An unresolved difficulty with this process is the fact that higher olefins exhibit limited miscibility with the aqueous phase containing the catalyst, resulting in low rates of reaction.

Hydroformylation occurs in the aqueous phase and product separation is therefore facile, thus allowing the hydroformylation of higher C_n olefins than propylene.

In principle, the problem of 'heterogenization' of a homogeneous catalyst has been successfully solved!

3.3 Kinetics and mechanism

Two principal approaches have been taken towards understanding the mechanism of the hydroformylation reaction:

- The development of rate equations based on experimental data which describe the effects of changes in variables such as metal concentration, pressure and temperature on reaction rate and product distributions. In the early stages this proved problematic because of the severe conditions of pressure and temperature used, which led to difficulties in the generation of reproducible data.
- The study of reactions of metal carbonyls with olefins, primarily at low pressures, in attempts to obtain evidence for plausible reaction intermediates.

It is only relatively recently that these approaches have converged, with the development of spectroscopic cells to observe reacting species directly under process conditions.

Kinetic information on hydroformylation is surprisingly sparse.

For Co-catalysed processes the simplified kinetic equation (Eqn 3.2) is now generally accepted to apply under commercial operating conditions:

$$\text{Rate} \propto [\text{olefin}][\text{Co}][P_{H2})[P_{CO}]^{-1} \quad (3.2)$$

For the phosphine-modified Rh catalysts the situation is more complex, with fractional dependencies on solvent and reactant under certain conditions, but a simplified approximation (Eqn 3.3) applies.

$$\text{Rate} \propto [\text{propylene}][\text{Rh}][P_{H2}] \quad (3.3)$$

The significant difference between the two sets of kinetic behaviour is the inverse dependence on P_{CO} for Co-catalysed reactions, which can be accounted for in terms of a requirement for the dissociation of CO at a key stage in the cycle, hence the name 'dissociative' pathway. Taken in conjunction with data from contemporary organometallic chemistry and in situ spectroscopic studies, in which most of the individual steps have been confirmed, the generally accepted reaction sequence for both Co processes is indicated in Eqns 3.4–3.10.

$$Co_2(CO)_6L_2 + H_2 \Longleftrightarrow 2HCo(CO)_3L \qquad L = CO, PR_3 \quad (3.4)$$

$$HCo(CO)_3L \Longleftrightarrow HCo(CO)_2L + CO \quad (3.5)$$

$$RCH{=}CH_2 + HCo(CO)_2L \Longleftrightarrow H(RCH{=}CH_2)Co(CO)_2L \quad (3.6)$$

A parallel sequence of reactions accounts for the formation of branched-chain products.

$$H(RCH{=}CH_2)Co(CO)_2L \Longleftrightarrow RCH_2CH_2Co(CO)_2L \quad (3.7)$$

$$RCH_2CH_2Co(CO)_2L + CO \Longleftrightarrow RCH_2CH_2Co(CO)_3L \quad (3.8)$$

$$RCH_2CH_2Co(CO)_3L \Longleftrightarrow RCH_2CH_2COCo(CO)_2L \quad (3.9)$$

$$RCH_2CH_2COCo(CO)_2L + H_2 \longrightarrow RCH_2CH_2CHO + HCo(CO)_2L \quad (3.10)$$

It is important to recognize that, with the exception of the hydrogenolysis step (Eqn 3.10), these reactions are reversible, as a consequence of which olefin, alkyl, and acyl isomerization can all occur. It should also be noted that there is a parallel sequence of reactions for the formation of branched-chain products.

The features of the catalytic cycle embody several of the concepts outlined in Section **2.1**, namely:

- activation of the catalyst by reaction of $Co_2(CO)_8$ or its bis-phosphine substituted derivative with H_2 to form the 18e cobalt hydridocarbonyl species $HCo(CO)_3L$ ($L = CO$ or PBu^n_3), Eqn 3.4,
- partial dissociation of CO to generate a coordinatively unsaturated 16e species $HCo(CO)_2L$ (Eqn 3.5),
- coordination of the olefin substrate (Eqn 3.6),
- formation of a 16e alkyl Co carbonyl species $RCH_2CH_2Co(CO)_2L$ (Eqn 3.7),
- coordination of CO to regenerate an 18e species (Eqn 3.8),
- migratory insertion to form the 16e acyl cobalt carbonyl $RCH_2CH_2COCo(CO)_2L$ (Eqn 3.9), and
- hydrogenolysis of the acyl cobalt–carbon bond to form the aldehyde with regeneration of the catalyst (Eqn 3.10).

The mechanistic nature of the final step is still unresolved. Two possibilities, hydrogenolysis by molecular hydrogen and bimolecular elimination via $HCo(CO)_xL$, may both occur at comparable rates with different metal carbonyl/phosphine catalyst combinations. However, the consensus view is that the former pathway (Eqn 3.10) predominates under commercial operating conditions.

Acyl hydrogenlysis is only believed to be rate determining in the hydroformylation of simple terminal olefins with unmodified Co catalysts; alkene addition to the metal hydride $HCo(CO)_2L$ becomes rate determining when more sterically hindered internal/branched olefins are used as substrates ($L = CO$), and with all alkenes when L = phosphine or phosphite.

An analogous mechanistic pathway can be developed for the phosphine-modified Rh catalyst system (Fig. 3.1).

H
Ph3P
Rh-CO
Ph3P
CO
CO
PPh3
H
CO
Ph3P
Rh
CO
R
H
PPh3
Rh
Ph3P
CO
RCH2CH2CHO
H
OC
Rh
Ph3P
CO
R
O
H
RCH2CH2C
H
Rh
Ph3P
PPh3
CO
OC
CO
Rh
Ph3P
CH2CH2R
H2
O
RCH2CH2C
PPh3
Rh
Ph3P
CO
PPh3
CO
OC
Rh-PPh3
Ph3P
CH2CH2R

Fig. 3.1 Catalytic cycle for the phosphine-modified rhodium hydroformylation catalyst.

Although the kinetic behaviour is open to interpretation in terms of an 'associative' rather than a 'dissociative' mechanistic pathway, it is generallyaccepted that the latter is operative under process conditions. Mechanistic details of the biphasic phosphine-modified Rh process are less well understood. Nevertheless, a generally unified overall picture of the reaction has emerged and it has been claimed that the hydroformylation reaction is now reasonably well understood in mechanistic terms. Even so, more than 60 years after its discovery, there are still aspects of the reaction mechanism that have not been clarified in detail.

3.4 Factors governing the choice of hydroformylation process

The four principal processes for the hydroformylation of olefins are in many respects complementary and each has its own niche.

It will be evident from the previous commentary that in many respects the commercial processes are complementary rather than competitive, and the principal factor governing the choice of a particular process concerns the versatility required in terms of substrate carbon chain length. Thus, if *n*-butyraldehyde is the required primary product, as an intermediate in the manufacture of 2-ethylhexanol, then the phosphine-modified Rh system is the clear process of choice. However, this process is of limited versatility and, if the requirement is for hydroformylation of a range of higher carbon number olefins to mixed aldehydes, then the original non-liganded Co process is satisfactory. If mixtures of alcohols are the desired products then the phosphine-modified Co process is appropriate, providing that some loss of olefin by hydrogenation to alkane can be tolerated.

Thus no process is ideal for all possible olefin conversions and, as is common in commercial processing, compromises have to be reached. Although the advantages of the phosphine-modified Rh processes in terms of mild operating conditions and high selectivities have led to their domination in the hydroformylation of propylene to *n*-butyraldehyde, it is significant that, for higher C_n alkenes, the use of Co-based processes currently outnumbers that of Rh by a factor of 9:1. Clearly, there is a trade-off, and a significant factor in this is no doubt the longevity of capital equipment installed many years ago for Co-based processes. This ratio is therefore set to decrease as new capacity comes on stream and the future trend will, where economically viable, be directed towards Rh-based catalysts and mild operating conditions.

3.5 Alkoxycarbonylation of olefins and acetylenes

Hydroformylation is only one member of a generic class of reactions which include the addition of CO and other co-reactants HX (where X

= H, OH, and alkoxy) to olefins (and acetylenes) with the formation of aldehydes, acids, and esters, respectively (Eqn 3.11), some of which are assuming increasing commercial significance.

In addition to hydroformylation, the closely related hydro- and alkoxy-carbonylation of alkenes and alkynes is assuming increasing commercial significance.

$$RCH{=}CH_2 + CO + HX \longrightarrow RCH_2CH_2COX \qquad (3.11)$$

Methoxycarbonylation of buta-1,3-diene to dimethyl adipate

BASF has recently developed a two-step Co-catalysed reaction for the *selective* conversion of buta-1,3-diene, CO and methanol to the dimethylester of adipic acid, Eqns 3.12 and 3.13; this is subsequently hydrolysed in a third stage to adipic acid (Eqn 3.14) for use as an intermediate in the manufacture of Nylon 6,6′ (see Chapter **5**).

The selective dimethoxycarbonylation of buta-1,3-diene has been developed into a process route for the manufacture of dimethyl adipate.

$$CH_2{=}CHCH{=}CH_2 + CO + CH_3OH \longrightarrow CH_3CH{=}CHCH_2CO_2CH_3 \qquad (3.12)$$

$$CH_3CH{=}CHCH_2CO_2CH_3 + CO + CH_3OH \longrightarrow CH_3CO_2(CH_2)_4CO_2CH_3 \qquad (3.13)$$

$$CH_3CO_2(CH_2)_4CO_2CH_3 + 2H_2O \longrightarrow HO_2C(CH_2)_4CO_2H \qquad (3.14)$$

The stepwise methoxycarbonylation involves the reaction of buta-1,3-diene and methanol with CO at *ca.* 600 bar and 120–140°C in the presence of relatively high concentrations of pre-formed $HCo(CO)_4$ and N-containing base such as pyridine to yield methyl-3-pentenoate in 90% selectivity (Eqn 3.12). The use of high catalyst concentrations ensures rapid carbonylation and minimizes typical side-reactions such as dimerization and oligomerization of buta-1,3-diene. In the second step (Eqn 3.13), some of the base is removed because it inhibits the methoxycarbonylation. Prior isomerization of 3- to 4-pentenoic acid ester is a prerequisite for the subsequent carbonylation to generate dimethyl adipate. To ensure internal double bond rearrangement, the reaction conditions are adjusted to 150 bar and 150–170°C to give the adipic ester in 80% selectivity. After hydrolysis of the ester, adipic acid is obtained with an overall selectivity of *ca.* 70%. Mechanistic details of the reaction are not clear and could involve either hydridocarbonyl (see Eqns 3.4–3.10) or methoxycarbonyl pathways (or both) (Fig. 3.2). In the latter scheme pyridine initially reacts with $Co_2(CO)_8$ to generate $[Co(py)_6][Co(CO)_4]_2$. Reaction with methanol yields the neutral methoxycarbonyl species $MeOCOCo(CO)_3$ and $HCo(CO)_4$. The former reacts with buta-1,3-diene to give the methoxycarbonylbutenyl complex which, in turn, reacts with $HCo(CO)_4$ to give methyl-3-pentenoate. It seems likely that the high

The role of the N-containing bases contrasts with that of the phosphorus ligands in phosphine-modified hydroformylation (Section **3.2**) in that they act as genuine catalyst promoters by accelerating overall reaction rate *and* enhancing selectivity (Section **1.2**) to the required dimethoxycarbonylation product isomer.

CO pressures used in the process are not directly related to the catalytic cycle but are required to stabilize the cobalt carbonyl intermediates in the liquid phase (Section **3.2**).

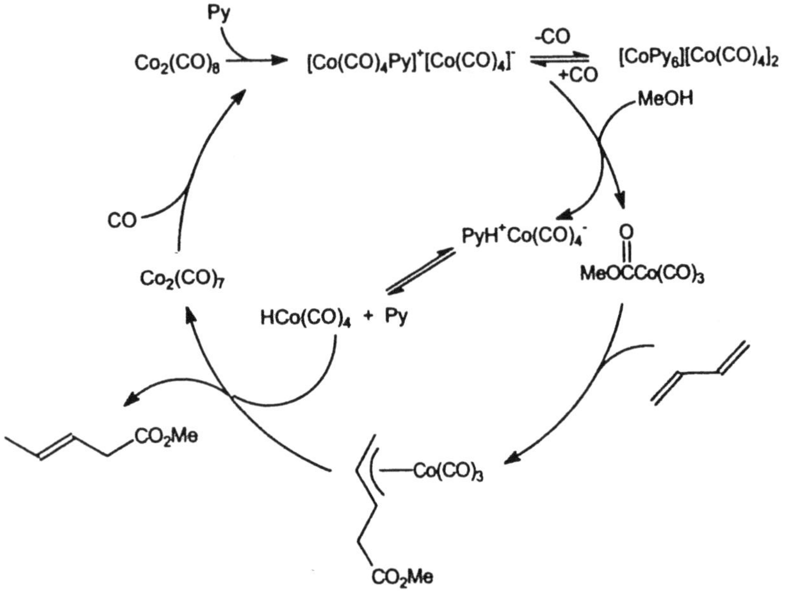

Fig. 3.2 The methoxycarbonyl pathway for cobalt-catalysed formation of esters.

Although plans were announced for the construction of a 60,000 te y^{-1} process it appears that the operation has not progressed beyond the pilot plant stage.

Methoxycarbonylation of methylacetylene to methyl methacrylate and of ethylene to methyl propionate

(Meth)acrylic acid and esters are large-volume industrial chemical intermediates used for the production of co- and homo-polymers. Of the methacrylics, methyl methacrylate (MMA) is by far the most important intermediate, used mainly in the manufacture of Perspex, with a production of *ca.* 1.5 mte y^{-1}. Conventional manufacture is by the stoichiometric Acetone Cyano Hydrin (ACH) process which, as a consequence of the production of *ca.* 2.4 tonnes of ammonium bisulphate/sulphuric acid waste per tonne of MMA, is faced with increasing environmental costs, and is therefore a prime candidate for the development of alternative production routes.

The carbonylation of alkynes has been known since the pioneering work of Reppe in the 1930s but, until recently, commercially viable routes to methacrylate derivatives have been hindered by insufficiently high catalytic activities and selectivities. A recent development in this area has been the discovery, by Shell, of a new class of highly efficient homogeneous Pd catalysts for the carbonylation of alkynes. These catalysts allow the selective and rapid carbonylation not only of acetylene but also of homologues such as methylacetylene (propyne) (Eqn 3.15), and do so with extremely high activity and selectivity to the *branched* unsaturated products, as required for the formation of methyl methacrylate. Thus, for the first time, they allow the development of a cost-effective MMA process based on carbonylation technology.

A new range of Pd catalysts has been developed for the selective methoxycarbonylation of methylacetylene to methyl methacrylate.

$$CH_3C{\equiv}CH + CO + CH_3OH \xrightarrow{[Pd]} CH_2{=}C(CH_3)COOCH_3 \qquad (3.15)$$

The essential feature of the catalyst systems is that they are formed from the combination of a Pd(II) species with (i) a ligand (L) containing a 2-pyridyl-phosphine moiety and (ii) a proton source containing counterions of weak coordination capacity.

Such catalysts are very efficient for the conversion of methylacetylene as the alkyne and methanol as the nucleophilic co-reagent. Typical product formation rates of *ca.* 20,000–50,000 mol mol^{-1} Pd h^{-1} are observed under mild reaction conditions (10–60 bar and 45-60°C). Highest reaction rates are observed at a L:Pd ratio of *ca.* 20 and selectivities are reported to approach 99.95%. Although details of the chemistry are currently open to debate, it has been proposed that the 2-pyridylphosphine ligand plays an essential role both as a bidentate chelating P⌒N ligand in the *selectivity*-determining step and as a unidentate ligand in the *rate*-determining step of the catalytic cycle.

Although this new range of catalytic systems has not yet been commercialized for methyl methacrylate manufacture, partly because of uncertainties in relation to the availability of the methylacetylene feedstock and required production capacities for MMA, this type of exciting new chemistry is of practical interest for at least two additional reasons. First, closely related cationic Pd-tertiary phosphine complexes containing weakly coordinating anions have been found to provide efficient catalysts for the methoxycarbonylation of ethylene to methyl propionate (Eqn 3.16), which itself comprises a useful

$$CH_2{=}CH_2 + CO + CH_3OH \xrightarrow{[Pd]} CH_3CH_2CO_2CH_3 \qquad (3.16)$$

intermediate in another potential production route to MMA. Second, when *monodentate* tertiary phosphine ligands are replaced by *bidentate* phosphines, there is a dramatic switch of selectivity in favour of the formation of polyketone – a regular 1:1 co-polymer of carbon monoxide and ethylene. A process for the production of polyketone based entirely on this new development in Pd chemistry has recently

A range of closely related Pd catalysts has been discovered for the both the selective methoxy-carbonylation of ethylene to methyl propionate and the formation of perfectly alternating co-polymers of ethylene and CO.

been commercialized by Shell (see Section **6.3**). Finally, the advent of this range of Pd catalysts is noteworthy in breaking the domination of Co, Rh and Ir catalysts in both carbonylation (see Chapter **4**) *and* hydroformylation chemistry.

3.6 Future prospects

Synthesis of fine chemicals

Future opportunities in hydroformylation seem likely to be offered by applications in the fine or speciality chemicals industry.

During the past 50 years the hydroformylation reaction has evolved from one which required relatively severe reaction conditions, effectively limiting the reaction to the manufacture of commodity chemicals, to one which can be operated under ambient conditions, thus bringing the reaction within the capability of the synthetic organic chemist. It seems likely therefore that new developments are most likely to appear in the fine chemicals industry. In fact it has already been used for a number of years in facilitating key steps in the BASF and Hoffmann-LaRoche syntheses of Vitamin A (see Section **7.2**).

An aspect of particular importance is asymmetric hydroformylation, which has been the subject of continuing research effort during the past 15–20 years. To date only moderate enantioselective excesses have been observed, probably as a consequence of the many reversible steps, outlined in Eqns 3.4–3.10 and Fig. 3.1, that can lead to isomerization and racemization. Refinement in the selection of ligands may enable these difficulties to be overcome.

Attempts to heterogenize hydroformylation catalysts

To date, attempts at the preparation and application of solid-supported heterogeneous hydroformylation catalysts, of the long-term stability required of commercial operation, have been thwarted because of slow loss, by leaching, of the catalyst into the liquid phase (*cf.* the discovery of hydroformylation).

Finally, there has been considerable interest over the years in the development of surface-supported or 'anchored' hydroformylation catalysts, with the twin aims of facilitating product separation from catalyst and reactants and of minimizing the problems of catalyst deposition onto reactor walls (Sections **1.2** and **2.5**). These attempts have, *in practice*, been singularly unsuccessful, in spite of numerous *claims* to the contrary, because of slow (ppm quantities) leaching of catalyst from the support into the liquid phase. Successful application of the biphasic approach to product separation has been demonstrated for both phosphine-modified Rh hydroformylation catalysts and Ni-catalysed oligomerization of ethylene to α-olefins (see Section **6.2**), and this approach appears to provide a more convenient and promising solution to the problem of product separation in such processes. Other emerging techniques such as fluorous biphase catalysis in combination with supercritical fluid processing *may* provide viable solutions to the problem of hydroformylation of higher C_n olefins with phosphine-modified Rh catalysts.

4 Acetic acid and acetyl chemicals

4.1 Background to manufacturing routes

The acetyl chemicals family includes acetic acid, acetaldehyde, methyl acetate, acetic anhydride and vinyl acetate, and the current capacity for their manufacture is *ca.*7 mte y^{-1} worldwide.

Acetic acid has been manufactured in large quantities for 150 years and is second only to methanol as a C_1-derived organic chemical. It is used in a wide range of end-products, e.g. vinyl acetate (50%), acetate esters (15%) (solvents), as a solvent specifically for the oxidation of *p*-xylene to terephthalic acid (10%), and in food and pharmaceutical applications. Vinyl acetate represents the largest and fastest growing outlet for acetic acid as a monomer or co-monomer in a variety of polymers that are major components of emulsion paints, adhesives and textiles. Acetic anhydride may be produced by the carbonylation of methyl acetate, the energy-intensive and relatively inefficient 'ketene' route, or by the oxidation of acetaldehyde. It is used mainly as an acetylating agent and in cellulose acetate manufacture (films and packaging materials).

Acetic acid is second only to methanol as a C_1 -derived chemical.

The principal applications of acetyl chemicals are as solvents (particularly for *p*-xylene oxidation) and vinyl acetate monomer for polymer manufacture.

Step changes in the processes used for the production of acetic acid and the acetyl chemicals, most of which involve homogeneous catalysis, or at least initiation, by transition metal ions or complexes, during this 150-year period reflect some of the underlying general trends in the chemical industry. These include the shift towards more energy-efficient processes, inherently safer routes (reductive vs. oxidative processes) and, as they have become readily available, more economic feedstocks. Currently, methanol carbonylation accounts for more than 60% of the world production of acetic acid.

Early production routes

In common with many other organic chemicals in the early days of the chemical industry, acetic acid was initially produced (in the 1850s) mainly by fermentation, together with the bacterial souring of wine. The distillation of wood was found to yield more concentrated solutions of acetic acid (15 to over 90%) and 10,000 te y^{-1} was still produced in the USA using this route in the mid-1960s.

The original production route (1850–1930) to acetic acid was based on fermentation.

Acetylene

The first major synthetic process was based on the hydration, or hydrolysis, of acetylene to acetaldehyde catalysed by Hg^{2+}, followed by oxidation to either acetic acid or acetic anhydride, Eqns 4.1 and 4.2.

Acetylene hydration was introduced in *ca.* 1930 and predominated acetyl chemicals manufacture until 1955–1960.

$$HC{\equiv}CH + H_2O \xrightarrow{Hg^{2+}} CH_3CHO \xrightarrow[O_2]{Mn^{2+}} CH_3CO_2H \quad (4.1)$$

$$CH_3CHO \xrightarrow[O_2]{Co^{2+},\ Cu^{2+}} (CH_3CO)_2O \quad (4.2)$$

Further reaction of acetaldehyde with acetic anhydride affords ethylidene diacetate which can be thermally de-acetylated to form vinyl acetate with the elimination of acetic acid (Eqn 4.3).

$$CH_3CHO + (CH_3CO)_2O \longrightarrow CH_3CH(OCOCH_3)_2 \xrightarrow{\Delta} CH_2{=}CHOCOCH_3 + CH_3CO_2H \quad (4.3)$$

Acetylene itself was manufactured from coke via the calcium carbide route, a highly energy-intensive process. Notwithstanding the disadvantages, namely that acetylene is hazardous and difficult to handle, and that Hg was lost in product recovery, thus making it environmentally unacceptable (certainly by present-day standards), this process was used extensively in Europe.

The use of acetylene as feedstock for the manufacture of acetyl chemicals proved to be energy-intensive and hazardous, and was environmentally unacceptable.

An alternative production route, particularly in the USA, was based on the oxidation of ethanol catalysed by Co and Cr acetates. In this process metal ions are involved in the initiation step, namely radical generation (Section **1.3**), but the oxidation reaction itself is believed to involve a radical chain mechanism. These two routes were the predominant processes used for the production of acetic acid for over 30 years, until the period 1955–1960, when two new major step-changes occurred, largely in parallel with the increased availability of oil-based feedstocks and the development of the petrochemical industry.

Paraffin oxidation

The first step-change involved the discovery of two alkane oxidation processes, namely the short-chain, e.g. butane, oxidation developed by Celanese in the USA, and the naphtha oxidation process discovered by BP in Europe. In the former a metal salt, usually Co acetate, is used as initiator to direct the cleavage of the alkane chain to give maximum yields of acetic acid (Eqn 4.4).

Butane and naphtha oxidation processes were developed in the early 1950s.

$$C_4H_{10} + {}^5/_2\,O_2 \xrightarrow{Co(OAc)_2} \underset{40\text{–}60\%}{2CH_3CO_2H} + H_2O \qquad (4.4)$$

This process is of relatively low selectivity and the economics are therefore heavily dependent on the sale of by-products.

Alkane oxidation is a radical chain process and the main propagation steps need not involve the metal ion (Section **1.3**). Cobalt is thought to participate in the initiation and decomposition of alkylhydroperoxide intermediates. Paraffin hydrocarbon oxidation processes are still operated today, and are competitive in situations where co-products can be marketed successfully, but in general have fallen out of favour relative to other newly developed technologies.

Alkane oxidation processes are generally of low selectivity and dependent on full utilization of by-products.

Ethylene oxidation

The second development during the 1950s was the discovery, by Wacker Chemie, of a simple high-yield selective oxidation of ethylene to acetaldehyde or, in the presence of acetic acid, to vinyl acetate, Eqns 4.5 and 4.6.

$$CH_2{=}CH_2 + {}^1/_2\,O_2 \longrightarrow CH_3CHO \longrightarrow CH_3CO_2H \qquad (4.5)$$

$$CH_2{=}CH_2 + {}^1/_2\,O_2 \xrightarrow{CH_3CO_2H} CH_2{=}CHOCOCH_3 \qquad (4.6)$$

The Pd-catalysed selective oxidation of ethylene to acetaldehyde was commercialized in 1960.

Methanol carbonylation

A subsequent development which now dominates all new processes for the production of acetic acid is the use of C_1 feedstocks, such as CH_3OH, and carbonylation-based technology. These homogeneously catalysed processes include a high-pressure version, using halide-promoted Co catalysts, developed by BASF in 1966, and the much lower pressure halide-promoted Rh- and Ir-based processes, introduced by Monsanto and BP Chemicals in 1970 and 1996, respectively.

Iodide-promoted Co-, Rh- and Ir-catalysed methanol carbonylation processes for the manufacture of acetic acid were introduced in 1966, 1970, and 1996, respectively.

Important subsequent extensions to Rh-based carbonylation technology have included a process for the carbonylation of methyl acetate to acetic anhydride (Tennessee Eastman/Halcon), which also comprises the first totally synthesis gas-based route to this important intermediate, and a co-carbonylation process for the conversion of methanol and methyl acetate to acetic acid and acetic anhydride respectively, developed by BP Chemicals in 1988.

A totally C_1 -based route to acetic anhydride (via methyl acetate carbonylation) was commercialized in 1983.

Trends

Over the past 150 years of acetyl chemicals manufacture the trend has been towards more economic, selective, safer and environmentally friendly processes.

The overall trends in the production of acetic acid and acetyl chemicals reflect a shift away from high-energy (and therefore usually expensive) intermediates such as acetylene, towards lower-energy materials (and successively more economic feedstocks) such as paraffins, olefins, methanol and synthesis gas.

Today, methanol carbonylation technology accounts for more than 60% of world acetic acid production.

Particularly important features in this context have been the Wacker process, which proved to be a major step in the displacement of acetylene as a starting material for the manufacture of organic chemicals, and the subsequent displacement of ethylene oxidation by C_1 feedstocks such as CH_3OH and the use of carbonylation-based technology. In common with hydroformylation (Chapter **3**) many of the recent developments have parallels with discoveries in the use of soluble metal complexes as catalysts, as illustrated in the following discussion of the oxidation of ethylene to acetaldehyde, the carbonylation of methanol to acetic acid, and of methyl acetate to acetic anhydride.

4.2 Oxidation of ethylene to acetaldehyde

It has been known since 1894 that ethylene can be stoichiometrically oxidized to acetaldehyde by aqueous $PdCl_2$ (Eqn 4.7).

$$CH_2{=}CH_2 + [PdCl_4]^{2-} + H_2O \longrightarrow CH_3CHO + Pd^0 + 2HCl + 2Cl^- \quad (4.7)$$

The stoichiometric Pd-mediated oxidation of C_2H_4 to CH_3CHO is made catalytic by the incorporation of a Cu-based redox system.

However, it was not until 1956 that it was discovered (by Wacker Chemie) that the reaction could be made catalytic (in Pd) by linking it to a Cu-based redox system, i.e. (i) re-oxidizing Pd(0) with $CuCl_2$, and (ii) subsequently re-oxidizing the resultant Cu(I) with molecular oxygen, Eqns 4.8 and 4.9.

$$2Cl^- + Pd^0 + 2CuCl_2 \longrightarrow [PdCl_4]^{2-} + 2CuCl \quad (4.8)$$

$$2CuCl + \tfrac{1}{2}O_2 + 2HCl \longrightarrow 2CuCl_2 + H_2O \quad (4.9)$$

In this way $CuCl_2$ acts as an essential promoter (or co-catalyst) of the Pd-catalysed hydration of ethylene to acetaldehyde and the *net* reaction from Eqns 4.7 – 4.9 becomes Eqn 4.10.

$$CH_2{=}CH_2 + \tfrac{1}{2}O_2 \longrightarrow CH_3CHO \quad (4.10)$$

In effect the Wacker process is stoichiometric with respect to Cu but catalytic in Pd.

Acetaldehyde is produced in yields and selectivities of *ca.* 95%. Minor by-products are 2-chloroethanol, ethyl chloride, acetic acid, chloroacetaldehydes and acetaldehyde condensation products. Such was the significance of this development that it was commercialized in

1960, only four years after its discovery.

In commercial practice, two methods of operation have been developed. In the original two-stage Wacker process, ethylene is stoichiometrically oxidized to acetaldehyde by a solution containing $PdCl_2$ and $CuCl_2$ in dilute HCl at 10 bar total pressure and 100–110 °C. After separation of acetaldehyde the aqueous solution containing $PdCl_2$ and CuCl is continuously circulated through a second reactor, in which the catalyst is regenerated by oxidation with air under similar reaction conditions, after which it is then recycled to the first stage.

The original Wacker process operates in two separate stages which are combined in the Hoechst variant. Each has attendant advantages and disadvantages.

In the single-stage variant developed by Hoechst the catalyst is regenerated in situ. An O_2/C_2H_4 mixture is fed to an aqueous solution of $PdCl_2$ and $CuCl_2$ at 3 bar pressure and 100–120°C. In both cases the heat produced by the reaction is used to remove the volatile acetaldehyde by fractionation without catalyst decomposition. Although the two-stage process is more capital intensive, it allows the use of air rather than oxygen, thus avoiding explosion hazards inherent in the mixing of oxygen and ethylene.

Ease of product separation was the key to successful operation of the process.

Remarkably, the Hoechst–Wacker Pd-catalysed oxidation of ethylene to acetaldehyde was commercialized only 4 years after its discovery.

Kinetics and mechanism

The kinetic equation for the reaction is given by Eqn 4.11.

$$-d[C_2H_4]/dt = k[PdCl_4^{2-}][C_2H_4]/[H_3O^+][Cl^-]^2 \quad (4.11)$$

The inverse square dependence on the chloride ion concentration may be explained in terms of two equilibria involving consecutive replacement of chloride ligands by (i) ethylene (Eqn 4.12), and (ii) water (Eqn 4.13), and which, together, determine the concentration of the aquopalladium complex $[(C_2H_4)PdCl_2(H_2O)]$.

$$[PdCl_4]^{2-} + C_2H_4 \Longleftrightarrow [(C_2H_4)PdCl_3]^- + Cl^- \quad (4.12)$$

$$[(C_2H_4)PdCl_3]^- + H_2O \Longleftrightarrow [(C_2H_4)PdCl_2(H_2O)] + Cl^- \quad (4.13)$$

Opinion is still divided as to how the next step occurs. One possibility concerns sequential dissociation of the acidic hydrogen (Eqn 4.14), which would explain the inhibition by acid, followed by rearrangment within the coordination sphere to give the β-hydroxyethyl complex (Eqn 4.15), the formation of which is rate-determining.

$$[(C_2H_4)PdCl_2(H_2O)] \Longleftrightarrow [(C_2H_4)PdCl_2(OH)]^- + H^+ \quad (4.14)$$

$$[(C_2H_4)PdCl_2(OH)]^- \longrightarrow [(HOCH_2CH_2)PdCl_2]^- \quad (4.15)$$

An alternative view is that nucleophilic attack on coordinated ethylene from outside the coordination sphere produces a four-coordinate hydroxyethyl compound which loses chloride in the rate-determining step (Eqn 4.16).

$$[(C_2H_4)PdCl_2(H_2O)] \xrightarrow{OH^-} [(HOCH_2CH_2)PdCl_2(H_2O)]^- \xrightarrow{-Cl^-} [(HOCH_2CH_2)PdCl(H_2O)] \quad (4.16)$$

In either case the hydroxyethyl compound rearranges and eliminates acetaldehyde with the formation of metallic palladium (Eqn 4.17), re-oxidation of which occurs according to Eqns 4.8 and 4.9.

$$[(HOCH_2CH_2)PdCl_2]^- \longrightarrow CH_3CHO + Pd^0 + HCl + Cl^- \quad (4.17)$$

This process represented the first practical demonstration of the economic viability of using a precious metal catalyst on a large scale.

No suitable heterogeneous catalyst equivalent has been found.

Pd-catalysed ethylene oxidation was not only the first industrial catalytic reaction to use transition metal–olefin chemistry but also the first demonstration that a precious metal could be used economically in a homogeneously catalysed reaction on an industrial scale. The process remains one of the most elegant applications of homogeneous catalytic oxidation and all attempts to develop a heterogeneous version have been unsuccessful. One reason why the homogeneous process has maintained its commercial competitiveness is the volatility of the product which lends itself to ready separation from the catalyst on a continuous basis without catalyst decomposition (Section **1.2**).

Vinyl acetate manufacture

When Pd-catalysed oxidation of ethylene is carried out in acetic acid rather than water the product is vinyl acetate, an important monomer. In 1960, as an extension to the original Hoechst–Wacker process, it was discovered that vinyl acetate could be produced by passing ethylene at 10 bar pressure into solutions of $PdCl_2$ in acetic acid containing Na or Li acetates at 120–130°C (Eqn 4.18).

$$CH_2{=}CH_2 + 2CH_3CO_2Na + [PdCl_4]^{2-} \longrightarrow CH_2{=}CHOCOCH_3 + CH_3CO_2H + 2NaCl + Pd^0 + 2Cl^- \quad (4.18)$$

Combination with the copper redox system, Eqns 4.8 and 4.9, facilitates catalyst regeneration, giving the *net* reaction in Eqn 4.19.

$$CH_2{=}CH_2 + CH_3CO_2H + {}^1/_2\,O_2 \longrightarrow CH_2{=}CHOCOCH_3 + H_2O \quad (4.19)$$

Overall yields in the range 90–95% based on both ethylene and acetic acid were claimed. This homogeneously catalysed process was commercialized, but later abandoned, because of severe corrosion problems associated with the use of acetic acid in such a strongly oxidizing environment. In this case attempts at developing an equivalent heterogeneous catalyst, based on the same chemical principle, were successful. Heterogeneous catalysts used in the gas-phase process include $PdCl_2/CuCl_2/C$ and $PdCl_2$/alumina. Vinyl acetate is produced in >90% selectivity at 5–10 bar and 160°C.

Corrosion problems associated with the practical operation of the homogeneous catalyst required the successful development of an equivalent heterogeneous catalyst for vinyl acetate manufacture.

4.3 Carbonylation of methanol to acetic acid and acetic anhydride

The formal insertion of CO into the C–O bond of methanol yields acetic acid (Eqn 4.20), which, in the form of its methyl ester, may undergo further carbonylation with the formation of acetic anhydride (Eqn 4.21).

$$CH_3OH + CO \xrightarrow{[M]} CH_3CO_2H \tag{4.20}$$

$$CH_3COOCH_3 + CO \xrightarrow{[M]} (CH_3CO)_2O \tag{4.21}$$

Both reactions are catalysed by metals [M] from the Co, Rh, and Ir triad, in the presence of halide, particularly iodide, promoters. They are closely interlinked by the final step of an organic reaction cycle which involves hydrolysis, methanolysis, or acetolysis of CH_3COI, with the formation of HI, and the corresponding acetyl derivatives (Eqn 4.22).

Iodide-promoted methanol carbonylation processes catalysed by Co, Rh, and Ir are now available.

$$CH_3COI + HOR \longrightarrow CH_3COOR + HI \quad (R = H, CH_3, COCH_3) \tag{4.22}$$

This step is the primary distinguishing feature of each carbonylation process. The HI liberated then reacts with methanol, methyl acetate (or dimethyl ether) to regenerate the methyl iodide promoter (Eqn 4.23).

$$HI + CH_3OR \longrightarrow CH_3I + HOR \quad (R = H, CH_3, COCH_3) \tag{4.23}$$

The principal function of the organometallic catalyst is to promote oxidative addition and migratory insertion, thus increasing the carbon number of the reacting organic substrate, e.g. CH_3I, by one (Eqn 4.24).

$$CH_3I + CO \xrightarrow{[M]} CH_3COI \tag{4.24}$$

The principal function of the metal catalyst in methanol carbonylation and related reactions is to promote oxidative addition and migratory insertion.

For the co-carbonylation of methyl acetate and methanol, the reaction of acetyl iodide with methyl acetate (or dimethyl ether) is also a key reaction step (Eqn 4.25).

$$CH_3COI + CH_3OR \longrightarrow CH_3I + CH_3CO_2R\ (R = CH_3, COCH_3) \quad (4.25)$$

Original methanol carbonylation processes

In 1966 BASF described a high-pressure process based on Reppe chemistry for the carbonylation of methanol to acetic acid using an iodide-promoted Co catalyst. This was rapidly followed in 1968 by Monsanto's announcement of the discovery of a low-pressure carbonylation using an iodide-promoted Rh catalyst, which achieved commercial status in 1970. A comparison between the two processes (Table 4.1) demonstrates similar trends and advantages associated with the substitution of Rh for Co in hydroformylation (Section **3.2**). The principal role played by iodide as promoter/co-catalyst in both systems is in the conversion of methanol to the more electrophilic methyl iodide (Eqn 4.26).

$$CH_3OH + HI \longrightarrow CH_3I + H_2O \quad (4.26)$$

Table 4.1 Comparison between Co- and Rh-based methanol carbonylation processes

Operating parameters	*Cobalt (BASF)*	*Rhodium (Monsanto)*
Pressure (bar)	500–700	30–40
Temperature (°C)	230	180
Metal concentration (M)	*ca.* 10^{-1}	*ca.* 10^{-3}
Promoter	I^-	I^-
Selectivity (%) based on methanol	*ca.* 90	>99
By-products	CH_4, CH_3CHO, C_2H_5OH, CO_2	None (detection limit 0.1%)

The Co-catalysed process is characterized by both lower activity and selectivity relative to Rh and considerably more severe reaction conditions.

The lower activity of Co, relative to Rh, requires higher operating temperatures for equivalent production rates. As a consequence, very high CO pressures are necessary to stabilize the Co carbonyl intermediates under reaction conditions (Section **3.2**). Not only is the selectivity of the higher-pressure process lower, as a consequence of greater by-product formation, but it is also sensitive to traces of H_2 (as impurities in the CO feed), whereas it has been claimed that the presence of up to 50% hydrogen is not detrimental to the Rh-based system. This represents a distinct process advantage in the context of the manufacture of pure CO from synthesis gas, in which residual traces of hydrogen are difficult and expensive to remove. The new (1996) Ir-catalysed 'Cativa' process developed by BP Chemicals is reported to operate under very similar reaction conditions to the original Monsanto process, although the standing concentrations of water, an important process parameter, are significantly different.

Kinetics and mechanism

A comparison between the kinetic behaviour of these processes, where the information is available (Table 4.2), highlights significant differences which are reflected in the nature of the rate-determining steps.

Table 4.2 Kinetics of methanol carbonylation processes based on cobalt, rhodium, and iridium

Reaction variable	*Effect on rate*		
Metal	Co	Rh	Ir
$[CH_3OH]$	first order	zero order	?
P_{CO}	second order	zero order	first order
$[I^-]$	first order	first order	$(\text{first order})^{-1}$
[M]	variable	first order	first order
$[CH_3CO_2H]$	?	zero order	?

The Co-catalysed reaction is dependent upon both substrate concentration and P_{CO} and the mechanism is thought to parallel that of the Co-catalysed hydroformylation reaction (Eqns 3.4–3.10), involving neutral rather than ionic species, with the addition of CH_3I (in place of olefin) to $HCo(CO)_4$, and its coordinatively unsaturated counterpart $HCo(CO)_3$, as the principal variant.

In contrast to Co, the kinetic equation for the Rh-catalysed reaction is remarkably simple (Eqn 4.27).

The Rh-catalysed system displays very simple kinetic behaviour.

$$-d[CH_3OH]/dt = k[Rh][I^-] \qquad (4.27)$$

Thus the rate is independent of the concentrations of both reactants and products; neither, therefore, has any kinetic influence.

The key intermediate in the reacting system is $[Rh(CO)_2I_2]^-$ and the rate-determining step in the process is oxidative addition of CH_3I to $[Rh(CO)_2I_2]^-$.

A key intermediate in this system is known to be $[Rh(CO)_2I_2]^-$, which is the principal species present in solution under operating conditions, according to in situ infrared spectroscopic measurements. The rate-determining step in the catalytic cycle is the oxidative addition of methyl iodide to $[Rh(CO)_2I_2]^-$ to give a $\{CH_3–Rh–\}$ complex which is very rapidly transformed into the corresponding acetyl species $\{CH_3C(O)–Rh–\}$ under CO. A binuclear rhodium acetyl iodo complex $[CH_3CORh(CO)I_3]_2^{2-}$, dimerized through a very weak iodide bridge, has been isolated from these reactions and structurally characterized. The overall catalytic cycle operates as summarized in Fig. 4.1.

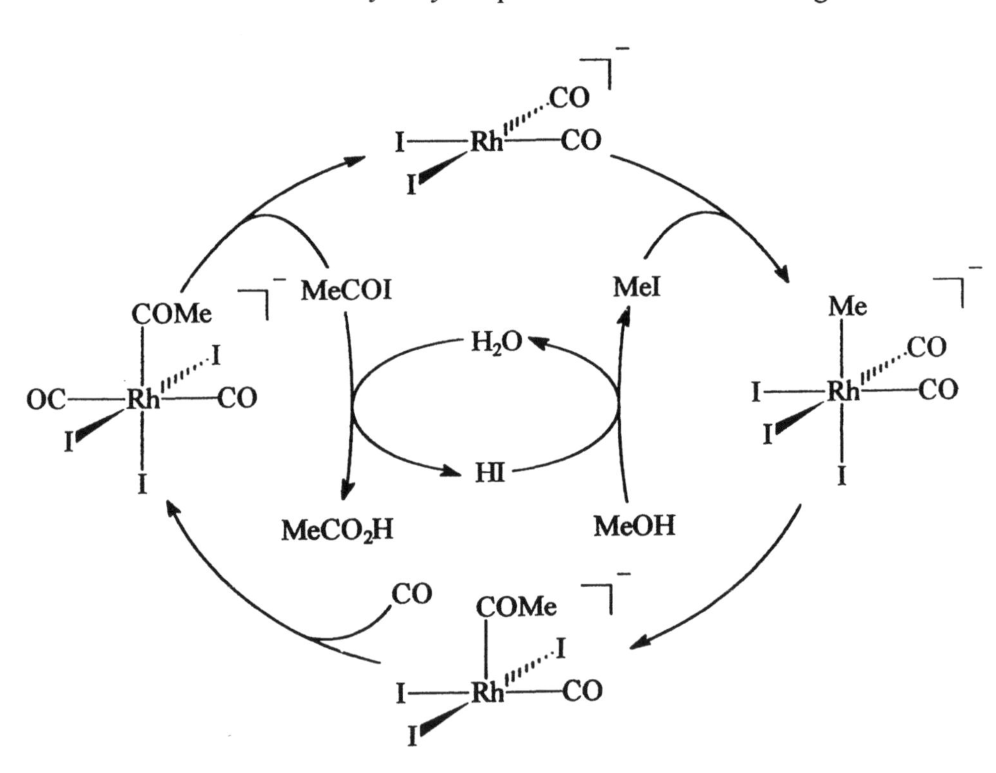

Fig. 4.1 Simplified reaction mechanism of the Monsanto process for the carbonylation of methanol to acetic acid.

The key to the high selectivity of the process is probably associated with the relative position of the rate-determining oxidative addition step and the subsequent cascade reaction sequence. Thus the extremely low standing concentration of the {CH_3–Rh–} species under reaction conditions makes it very unlikely that, in the presence of H_2, methane formation would comprise a serious competing reaction.

Although highly selective and efficient, under process conditions Rh-catalysed carbonylation does suffer from disadvantageous side-reactions. These include water gas shift chemistry, which represents loss of selectivity with respect to the CO raw material, and competitive oxidative addition of HI to $[Rh(CO)_2I_2]^-$. The latter, although only of minor significance relative to oxidative addition of CH_3I, can lead to loss of catalyst via the formation of $[Rh(CO)_2I_4]^-$ (Eqn 4.28), $[Rh(CO)I_4]^-$ and the ultimate precipitation of inactive and insoluble RhI_3, which is very difficult to regenerate.

Water gas shift chemistry and the oxidative addition of HI to $[Rh(CO)_2I_2]^-$ comprise minor but undesirable competing side-reactions.

$$[Rh(CO)_2I_2]^- + 2HI \longrightarrow [Rh(CO)_2I_4]^- + H_2 \qquad (4.28)$$

Finally, it is significant that a heterogeneous Rh catalyst, developed by Monsanto in parallel with the homogeneous process, displays analogous kinetic behaviour. However, the heterogeneous analogue was characterized by significantly reduced activity and slow loss of precious metal from the support (*cf.* Section **3.6**, attempts at heterogenizing hydroformylation catalysts), and, in the final analysis, the homogeneous process was preferred.

A heterogeneous version of Monsanto chemistry was developed in parallel with the homogeneous process, but ultimately the latter was preferred.

4.4 Developments to the original Monsanto process

As with hydroformylation (Chapter **3**), the catalytic activity of metals other than the Co, Rh, and Ir triad towards methanol carbonylation has been explored, but with very limited success. Although Ni catalysts have been demonstrated to provide high selectivities to acetic acid, the extent of side-reactions such as hydrocarbon formation has proved prohibitively significant. Hence, although the original BASF process remains in operation today, all subsequent developments and process improvements in methanol carbonylation have been based on Monsanto-type technology which clearly has significant process advantages, dominates all production capacity, and now accounts for approximately 60% of acetic acid manufacture.

Success in catalytic methanol carbonylation with metals other than Co, Rh, and Ir has been very limited.

Subsequent developments and refinements have encompassed several distinct, but closely related, process variations (Table 4.3). These have been directed towards optimization of the manufacture of either acetic acid or acetic anhydride, together with the co-production of both.

Monsanto-based carbonylation technology dominates all new production capacity for acetic acid.

Table 4.3 Development of carbonylation-based processes for the production of acetic acid and acetic anhydride

Company	*Date*	*Product*	*Metal*	*Co-catalyst*
BASF	1966	CH_3CO_2H	Co	HI
Monsanto	1970	CH_3CO_2H	Rh	MeI/HI
Hoechst Celanese	early 1980s	CH_3CO_2H	Rh	MeI/LiI
Tennessee Eastman	1983	$(CH_3CO)_2O$	Rh	MeI/LiI + $[R_4P]^+$ salts
Hoechst	mid-1980s	$(CH_3CO)_2O$	Rh	MeI + $[R_4P]I$
BP Chemicals	1988	CH_3CO_2H and $(CH_3CO)_2O$	Rh	MeI + imidazolium iodides, $ZrO(OAc)_2$
BP Chemicals	1996	CH_3CO_2H	Ir	MeI + Ru complexes

Acetic acid – the Hoechst Celanese 'low water' process

The first significant modification, the 'low water' process introduced by Hoechst Celanese, directly addressed one of the problems inherent in the original process, namely the requirement for a relatively high water content (typically 15 wt%) in the reacting system. This was necessary to maintain catalyst activity, to achieve economically viable production rates, and to maintain good catalyst stability. Because of the high water content in the reactor, the separation of acetic acid represents both a major energy usage and a capital cost penalty in the process. As a consequence much research effort has been directed towards devising modifications which allow operation at low water (*ca.* 5 wt%) concentrations, and which, at the same time, compensate for both the inherent decrease in reaction rate and catalyst stability under typical Monsanto reaction conditions.

The 'low water' variation was introduced to reduce the requirement for relatively high levels of water in the original Monsanto process. Operation under such conditions entails a difficult and expensive distillation step.

The Hoechst Celanese modification addressed this problem by the incorporation of an inorganic iodide promoter, LiI, at relatively high concentrations and above an I^- level not previously thought to be effective as a catalyst stabilizer and promoter. This has enabled the development of a 'low water' variant of the original Monsanto technology. Benefits of the use of LiI include (i) stabilization of the Rh catalyst at low water concentrations by preventing precipitation as insoluble RhI_3 (Section **4.3**), (ii) acting as an additional promoter of catalytic activity, and (iii) effectively optimizing $[Rh(CO)_2I_2]^-$ concentrations at low water levels. It has been proposed that the additional promotional role of inorganic iodide is associated with the

The 'low water' process variant uses LiI as an additional, ionic, iodide promoter of the reaction.

formation of an as yet unidentified highly nucleophilic five-coordinate dianionic intermediate $[Rh(CO)_2I_2L]^{2-}$ ($L = CH_3CO_2^-$) which is more active towards oxidative addition of methyl iodide than $[Rh(CO)_2I_2]^-$.

Acetic anhydride – carbonylation of methyl acetate

The second significant modification was the development, by Tennessee Eastman, of the carbonylation of methyl acetate to acetic anhydride, which also comprised the first example of a totally integrated synthesis gas-based process for such chemistry, dependent entirely on coal as feedstock (Eqn 4.29).

This process for acetic anhydride manufacture represents the first example of an integrated synthesis gas process based entirely on coal as feedstock.

$$\text{Coal} \longrightarrow CO/H_2 \longrightarrow CH_3OH \longrightarrow CH_3CO_2CH_3 \xrightarrow{CO} (CH_3CO)_2O \quad (4.29)$$

This 227,000 te y^{-1} capacity process came on stream at Kingsport, Tennessee, in October 1983 and a closely related process was completed by Hoechst in the mid-1980s.

The Tennessee Eastman plant is built on the site of a coal mine. Good-quality coal is gasified to CO/H_2 which in turn is converted into methanol. The process then operates in two stages, namely the reaction of methanol with acetic acid to give methyl acetate, followed by carbonylation to acetic anhydride. Both processes are highly selective and the overall selectivity is believed to be in excess of 96%, based on methanol. The acetic anhydride produced is used internally and is converted, with cellulose, into cellulose acetate for applications as photographic film bases, textile fibres, and cigarette filters. A by-product of the cellulose esterification is acetic acid, which is utilized in the esterification of methanol produced separately from CO/H_2.

As with the Monsanto system an iodide-promoted Rh catalyst is used at operating conditions of 50 bar (CO containing some H_2) and 190°C. The presence of H_2 is required primarily to reduce $[Rh(CO)_2I_4]^-$ to $[Rh(CO)_2I_2]^-$ and to maintain to a minimum the presence of unreactive Rh(III) species which otherwise accumulate at the low water conditions required for acetic anhydride production (Section **4.3**). In order to enhance the intrinsic catalytic activity, which is lower than under typical Monsanto conditions, catalytic promoters such as LiI are also employed together with co-catalysts such as quaternary phosphonium iodides.

The organometallic chemistry of methyl acetate carbonylation is closely related to that previously described for the carbonylation of methanol with a rate-determining oxidative addition of CH_3I to Rh and rapid migratory insertion comprising the key steps (see Fig. 4.2). For methyl acetate carbonylation, methanol is replaced by methyl acetate (or dimethyl ether) in the organic part of the reaction cycle and acetate is the leaving group rather than hydroxyl.

In the presence of additional co-promoters equivalent rates of methyl acetate carbonylation to those of methanol carbonylation are achieved.

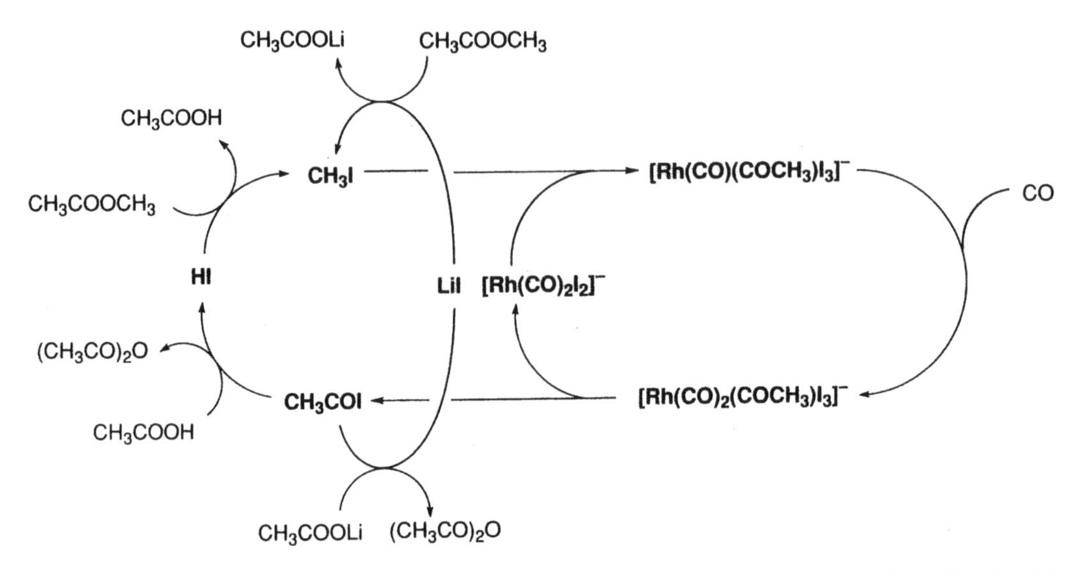

Fig. 4.2 Reaction mechanism for the carbonylation of methyl acetate to acetic anhydride. (Reprinted with permission from Wiley-VCH).

The role of Li is thought to be related to promotion of the ionic portion of the organic cycle. Here, LiI rather than HI reacts with methyl acetate to yield methyl iodide which is carbonylated. Lithium acetate then reacts rapidly with the acetyl iodide to give the anhydride. This effectively decreases the steady-state concentration of acetyl iodide, the presence of which can partially poison the Rh catalyst through the formation of molecular iodine. In contrast to methanol carbonylation, therefore, generation of methyl iodide via HI constitutes the minor reaction pathway.

The development of this process provides a good example of the trend away from oil-based feedstocks towards coal or natural gas.

Certain economic factors particular to Tennessee Eastman, for example substantial local deposits of high-grade coal, internal use of product acetic anhydride, recycling of acetic acid, etc., make the process particularly favourable to these operators, and the production of acetic anhydride by the carbonylation of methyl acetate provides another example of the trend away from oil-based feedstocks towards coal and/or natural gas.

Vinyl acetate

As a potential extension of this technology, Halcon, amongst others, has also been active in the development of homogeneously catalysed synthesis gas-based routes for the production of other members of the acetyl chemicals family; in particular, vinyl acetate.

Hydrocarbonylation of methyl acetate with CO/H_2 can produce ethylidene diacetate, which, after thermal cracking over a metal oxide catalyst, yields vinyl acetate and acetic acid, Eqns 4.30 – 4.33.

A production route to vinyl acetate is also accessible via synthesis gas technology.

$$2CH_3CO_2CH_3 + 2CO + H_2 \xrightarrow{[Rh]} CH_3CH(OCOCH_3)_2 + CH_3CO_2H \quad (4.30)$$

$$CH_3CH(OCOCH_3)_2 \xrightarrow{\Delta} CH_2{=}CHOCOCH_3 + CH_3CO_2H \quad (4.31)$$

$$2CH_3OH + 2CH_3CO_2H \longrightarrow 2CH_3CO_2CH_3 + 2H_2O \quad (4.32)$$

Net reaction:

$$2CH_3OH + 2CO + H_2 \longrightarrow CH_2{=}CHCO_2CH_3 + 2H_2O \quad (4.33)$$

The co-produced acetic acid may be recycled by reaction with methanol to provide methyl acetate feedstock. The reaction is catalysed by both Rh and Pd in the presence of iodide and N-base promoters (pyridine) under rather higher pressures than those used for methyl acetate carbonylation, namely, 140 bar CO/H_2 (1:2) and 150°C. Selectivities of up to 50% ethylidene diacetate have been claimed.

Nevertheless, in an age of wildly fluctuating oil prices (from below US \$10 per barrel in early 1999 to as high as US \$50 in late 2004), such a synthesis gas based-route requires considerable development before it can compete with existing technology (Section **4.2**).

Co-production of acetic acid and acetic anhydride

The bulk of the original Monsanto technology was purchased by BP Chemicals in 1986, and since then further process improvements have been made, in particular the development of a mixed co-product process in 1988. The basic concept of this operation is to co-produce acetic acid and acetic anhydride from methanol and CO as the only raw materials and to generate methyl acetate within the process, thus allowing the flexibility to tune the acetic acid/acetic anhydride product mixes to match market requirements.

The co-production of acetic acid and anhydride from CH_3OH and CO as the only raw materials allows much greater flexibility to tune the product mix to match market requirements.

The course of the co-carbonylation of methanol and methyl acetate can be considered to occur in three sequential stages, involving water formation (Eqn 4.34), and water uptake (Eqn 4.35),

$$2CH_3OH + CO \longrightarrow CH_3CO_2CH_3 + H_2O \quad (4.34)$$

$$CH_3CO_2CH_3 + H_2O + CH_3OH + 2CO \longrightarrow 3CH_3CO_2H \quad (4.35)$$

and, finally, at zero water and constant acetic acid concentrations, the direct carbonylation of methyl acetate to acetic anhydride is initiated (Eqn 4.36).

$$CH_3CO_2CH_3 + CO \longrightarrow (CH_3CO)_2O \quad (4.36)$$

N,N-dimethylimidazolinium iodide promoters, typically with $ZrO(OAc)_2$ as co-catalyst, are used in this Rh/I^--based process at 30 bar and 183°C. The role of the promoters appears to be associated with stabilization of the Rh catalyst system and to accelerate the rate of carbonylation.

Iridium catalysis

Finally, as referred to briefly in Section **4.3**, a remarkable step-change to existing technology has been the introduction, in 1996, of the 'Cativa' technology by BP Chemicals (now BP Amoco), incorporating the first commercial use of Ir, rather than Rh, as a catalyst for methanol carbonylation. The process gives improvements via much higher reactivity coupled with lower by-product formation and lower energy requirements for the purification of the product acid.

This development, which is claimed to be the most significant breakthrough in the industry in 25 years, represents the first commercial use of iridium as a carbonylation catalyst. The technology has been incorporated into two existing plants, in the USA and Korea, and in a new 500,000 te y^{-1} capacity plant currently under construction in Malaysia.

Although the catalytic activity of Ir in methanol carbonylation has been known for many years, it is characterized by both lower activity and selectivity than Rh *under typical Monsanto process conditions*. This is partly a consequence of the operation of a more complex catalytic cycle which involves both neutral and anionic iridium carbonyl halides as intermediates. However, under *modified* reaction conditions, particularly the 'low water' conditions preferred in the more recent technology, Ir catalysts have proved not only *significantly more stable* than their Rh counterparts but also *more active*.

Ir catalysts are considerably more stable than Rh under the preferred 'low water' operating conditions and are also more active.

At operating pressures and temperatures similar to those of the original Rh-catalysed process, a particularly significant feature of the new development is the high stability of the Ir catalyst at water levels as low as 0.5 wt%, which, by reducing distillation requirements, provides a highly desirable operating process advantage (see Hoechst Celanese development above). Significant kinetic differences between Rh and Ir catalysts have also emerged (Table 4.2). Oxidative addition of CH_3I is very rapid with Ir and the subsequent migratory insertion reaction now becomes rate determining, in complete contradistinction to the corresponding Rh chemistry. Another fundamental difference is that whereas high levels of CH_3I are required to give the highest reaction rates in Rh systems, with Ir the rate is largely independent of $[CH_3I]$. The inverse dependence on ionic iodide concentration suggests that very high reaction rates should be attainable by operating at low total iodide concentrations. It also implies that the inclusion of a species capable of assisting iodide abstraction could enhance the new rate-limiting step. Catalyst promoters containing Ru, Os, and Re have been found to be effective in this respect. In addition, they have been shown to act as CO donors, which has the *net* effect of decreasing the P_{CO} dependency of the unpromoted system.

The reaction rate law is of the form: $-d[CH_3OH]/dt = k[Ir][P_{co}]/[I^-]$.

By using co-promoters such as Ru complexes it has proved possible to increase the reaction rate and to effectively remove the dependence of rate on P_{co}.

The very high activity and robust nature of iridium catalysts at low water concentrations (productivities as high as 45 mol l^{-1} h^{-1} at 5 wt% water have been quoted), combined with the inherent stability of the catalyst under such conditions, leads to significant economic improvements over the Rh-based process (typical activities of which are *ca.* 10–15 mol l^{-1} h^{-1}). Also, in commercial operation, the formation of by-products such as propionic acid (via acetaldehyde), is substantially reduced.

4.5 Future prospects

Promoters

The role(s) of a range of promoters/co-catalysts of these Rh- and Ir-catalysed carbonylation reactions, although as yet incompletely understood, has allowed the development of several important process variations of the pioneering Monsanto technology. Further process improvements, incorporating other co-promoters, must be considered highly likely.

Heterogenization

Attempts at the 'heterogenization' of methanol carbonylation catalysts have been uniformly unsuccessful, despite many initial claims to the contrary (*cf.* hydroformylation, Chapter **3**) and the prospects for a successful solution to this problem must be considered remote.

Feedstocks

In this chapter we have seen the progression of step-changes in both the processes and feedstocks used for the manufacture of acetic acid and the acetyl chemicals during the past 150 years. What of the future?

The next step-change in the acetic acid saga seems most likely to emerge in the form of a direct route from either CH_4 or C_2H_6, e.g. the oxidative carbonylation of methane to methanol and acetic acid (Eqns 4.37 and 4.38), which has been recently demonstrated over Rh catalysts (at extremely low conversions!), or the selective oxidation of ethane (Eqn 4.39).

Perhaps the next step-change in the production of acetic acid will be derived from the oxidative carbonylation of CH_4.

$$CH_4 + CO + O_2 \xrightarrow{[Rh]} CH_3OH + CO_2 \quad (4.37)$$

$$CH_4 + 2CO + O_2 \xrightarrow{[Rh]} CH_3CO_2H + CO_2 \quad (4.38)$$

$$C_2H_6 + {}^3/_2\, O_2 \longrightarrow CH_3CO_2H + H_2O \quad (4.39)$$

It is hoped that a succeeding primer will provide the details of new processes based on these reactions!

5 Nylon intermediates: buta-1,3-diene hydrocyanation

5.1 Background

Co-monomers for Nylon 6,6′ production are adipic acid and hexamethylenediamine (HMD), both of which are traditionally derived from benzene.

Benzene has been used traditionally as the primary hydrocarbon source for the manufacture of Nylon, a polyamide, via the intermediates adipic acid and hexamethylenediamine (HMD). Adipic acid is manufactured first, a portion retained for incorporation into the final polyamide and the remainder converted to HMD, the amine co-monomer. Condensation polymerization (Eqn 5.1) of these co-monomers gives Nylon 6,6′, the most common form of Nylon. The numbering scheme is derived from the number of carbon atoms in each of the acid and amine components.

$$HO_2C\text{-}(CH_2)_4\text{-}CO_2H + H_2N\text{-}(CH_2)_6\text{-}NH_2 \longrightarrow$$

$$-[CO(CH_2)_4\text{-}CONH(CH_2)_6\text{-}NH]_n- + 2nH_2O \qquad (5.1)$$

Both adipic acid and HMD production routes are multistage processes.

The production route involves a complex sequence of steps in which benzene is first reduced to cyclohexane, oxidized to a cyclohexanone/ cyclohexanol mixture (called KA, *K*etone *A*lcohol), which is then oxidized further to adipic acid. The oxidation of cyclohexane to KA is carried out in the liquid phase with Co or Mn naphthenate catalysts. Oxidation occurs by a radical ion process in which the transition metal ions act as radical initiators (Section **1.3**). Conversions are limited to <10% in order to maximize selectivity to KA. The unseparated KA mixture is then further oxidized with nitric acid over a V/Cu catalyst to give adipic acid in 96% selectivity. An alternative to the corrosive acid oxidant, utilizing Cu and Mn acetate catalysts, has also been commercialized. Part of the purified adipic acid is set aside for nylon manufacture as acid co-monomer, and the remainder is converted, first into adiponitrile, and then HMD. This involves four process stages, namely amination of adipic acid to diammonium adipate, catalysed by boron phosphate in the vapour phase at 350°C, followed by two sequential dehydration steps, giving adipic acid diamide and finally adiponitrile. Overall selectivities of up to 90% adiponitrile are feasible.

The conversion of adipic acid into HMD involves four stages.

Adiponitrile is then purified before final conversion, by hydrogenation, to HMD (95%) over a heterogeneous Fe catalyst at 300 bar pressure and 150°C. Adipic acid and HMD are combined in water or methanol solution, from which a pure 1:1 adduct (Nylon salt) may be crystallized. The condensation polymerization is then carried out by heating the salt to 250–270°C, with continual removal of the water formed, to give Nylon 6,6′.

Alternative feedstocks for Nylon manufacture

The multi-step sequence of reactions required for the manufacture of Nylon from benzene has provided a commercial incentive to introduce step changes by the development of simpler routes based on alternative feedstocks, particularly for the production of adiponitrile. The C_3 and C_4 hydrocarbons propylene and buta-1,3-diene comprise the most obvious alternative building blocks. Indeed, recent developments at BASF have demonstrated a viable alternative production route to adipic acid via the selective dicarbonylation of buta-1,3-diene (Section **3.5**).

The C_3 and C_4 hydrocarbons propylene and buta-1,3-diene provide alternative feedstocks to benzene for the intermediates used in Nylon manufacture.

Propylene

Propylene is readily converted into acrylonitrile by ammoxidation using well-known mixed oxide (Bi/Mo) heterogeneous catalysts (Eqn 5.2).

$$CH_3CH{=}CH_2 + NH_3 + {}^3/_2\,O_2 \longrightarrow CH_2{=}CHCN + 3H_2O \qquad (5.2)$$

Ammoxidation of propylene gives acrylonitrile.

The acrylonitrile thus formed may be dimerized electrochemically to adiponitrile using the Monsanto electrohydrodimerization process, in which the overall cathodic reaction is represented by Eqn 5.3.

$$2CH_2{=}CHCN + 2e^- + 2H^+ \longrightarrow NC(CH_2)_4CN \qquad (5.3)$$

Electrohydrodimerization of acrylonitrile gives adiponitrile.

Buta-1,3-diene

Buta-1,3-diene also provides a viable feedstock for adiponitrile manufacture. The original three-stage indirect hydrocyanation process, developed by DuPont, involved chlorination of buta-1,3-diene to a mixture of chlorobutenes, followed by hydrocyanation to adiponitrile. However, overall reaction selectivity was poor, the process required the disposal of large quantities of NaCl, and improved technology was required.

The original DuPont process involved indirect hydrocyanation via chlorobutenes.

5.2 Hydrocyanation of buta-1,3-diene

The *direct* hydrocyanation of butadiene to adiponitrile, subsequently developed by DuPont, is based on the Ni-catalysed double *anti*-Markovnikov addition of HCN to butadiene (Eqn 5.4).

$$CH_2{=}CHCH{=}CH_2 + 2HCN \longrightarrow NC(CH_2)_4CN \quad (5.4)$$

Ni-catalysed hydrocyanation is a three-stage homogeneously catalysed process for which there is no heterogeneous counterpart.

The process was developed during the 1960s and first commercialized in 1971. It is now operated at three locations, Orange (Texas), Victoria (Texas) – the latter a conversion in 1983 from a plant which used the indirect hydrocyanation route via chlorobutenes – and in France in a Rhone Poulenc/DuPont joint venture, representing a total world capacity approaching 1.0 mte y^{-1}.

This homogeneously catalysed process, for which there is no known heterogeneous counterpart (Section **1.2**), is carried out in three stages:

- addition of the first equivalent of HCN,
- double-bond isomerization, and
- concurrent further isomerization and addition of the second equivalent of HCN.

An important feature of the process is the mild operating conditions which are used.

Each of these is operated under mild conditions of pressure and temperature, relative to those detailed in Chapters **3** and **4**, and involves catalysis by air- and moisture-sensitive nickel phosphite complexes of the type NiL_4.

First stage

This involves the production of mononitriles by the addition of one molecule of HCN to buta-1,3-diene catalysed by a Ni–aryl phosphite complex, in the presence of excess ligand, and produces a 70:30 mixture of C_5 mononitriles (Eqn 5.5). These comprise the desired linear 1,4-addition product, 3-pentenenitrile (containing a minor amount of 4-pentenenitrile), and the undesired branched product of 1,2-addition, 2-methyl-3-butenenitrile.

Addition of the first molecule of HCN gives a 2:1 mixture of linear and branched C_5 mononitriles.

$$CH_2{=}CHCH{=}CH_2 + HCN \xrightarrow[>70^{\circ}C]{NiL_4/L} CH_3CH{=}CHCH_2CN + CH_2{=}CHCH(CN)CH_3 \quad (5.5)$$

The reaction is operated above 70°C and at a modest pressure sufficient to ensure that unreacted buta-1,3-diene remains condensed. The isomers are separated by distillation and the branched product circulated to the second stage.

Second stage

In this stage the undesired 2-methyl-3-butenenitrile is isomerized into predominantly 3-pentenenitrile using a similar Ni catalyst, to which has been added a Lewis acid promoter such as $ZnCl_2$ (Eqn 5.6).

Isomerization of the branched to linear mononitrile occurs in the presence of a Lewis acid catalyst.

$$CH_2{=}CHCH(CN)CH_3 \underset{120^\circ C}{\overset{NiL_4/ZnCl_2}{\Longleftrightarrow}} CH_2{=}CHCH_2CH_2CN + CH_3CH{=}CHCH_2CN \quad (5.6)$$

Fortunately, 3-pentenenitrile is favoured thermodynamically over 2-methyl-3-butenenitrile by a factor of about 9 and isomerization occurs readily, at slightly higher temperatures than those used in the first stage.

Favourable thermodynamics prevail for the isomerization of the undesired isomer to 3-pentenenitrile.

Third stage

The third and final stage involves both internal double-bond migration to the more reactive 4-pentenenitrile and selective *anti*-Markovnikov addition of the second molecule of HCN with a similar catalyst system (Eqn 5.7).

$$CH_2{=}CHCH_2CH_2CN + CH_3CH{=}CHCH_2CN + HCN \xrightarrow[ZnCl_2,\ 80^\circ C]{NiL_4/L} NC(CH_2)_4CN \quad (5.7)$$

A combination of isomerization to terminal pentenenitrile and addition of a second molecule of HCN is also promoted by the presence of Lewis acids.

At this point it is fortuitous that 3-pentenenitrile isomerizes to 4-pentenenitrile more rapidly than to the thermodynamically preferred 2-pentenenitrile (by a factor as high as 70), since the conjugated isomer is believed to be a catalyst inhibitor. This provides a good example of *kinetic* reaction control. Temperatures required for the second addition of HCN are lower than for the first and practical conversion rates can in fact be achieved under ambient conditions.

Kinetic control favours the formation of 4-pentenenitrile over the thermodynamically more stable conjugated 2-isomer, prior to the second addition of HCN .

The use of $ZnCl_2$ as the Lewis acid is reported to give adiponitrile in *ca*. 83% overall selectivity at 80°C. The subsequent replacement of $ZnCl_2$ by BPh_3, a Lewis acid of greater steric bulk, has provided a significant improvement in selectivity (to >95% at 99% conversion) in the third stage at lower operating temperatures (40–50°C). The only major by-product of the process is 2-methylglutaronitrile (15%), together with a small amount of ethylsuccinonitrile (2%).

5.3 Reaction mechanism

Because of the mild operating conditions and the use of P-containing ligands, this process has proved particularly amenable to kinetic and mechanistic studies.

The chemical nature of this catalyst sytem, and the relatively mild reaction conditions under which it is effective, have made the process particularly amenable to detailed kinetic and mechanistic studies. As a consequence, a number of the postulated intermediate species involved in the individual steps of the process have been unequivocally identified from a combination of kinetic and ^{1}H, ^{13}C, and ^{31}P NMR spectroscopic measurements on either the actual or closely related organonickel complexes which comprise intermediates in the catalytic cycle.

Role of the ligand

The role of the ligand is critical. Phosphites are essential for high catalytic activity, and phosphites containing bulky substituents are essential for high selectivity to linear products. The use of phosphines rather than phosphites results in oxidation of Ni(0) to Ni(II) with catalyst deactivation.

In these transformations the nature and role of the phosphorus ligands are very important and the hydrocyanation process is a classic example in which the catalyst has been fine-tuned, by changing the substituents attached to phosphorus, to meet selectivity requirements. Among phosphorus compounds, only phosphites produce catalysts of sustained activity. If, for example, the more commonly available range of phosphine ligands is utilized (*cf.* hydroformylation chemistry, Chapter **3**), the outcome is oxidation of the nickel centre with the formation of Ni(II) cyanide complexes of the type $L_2Ni(CN)_2$, and consequent catalyst deactivation. Aryl phosphites are in fact necessary to maintain Ni in the zerovalent state during key stages of the catalytic cycle. Among the phosphites, only the relatively bulky triaryl phosphites with strongly electron-withdrawing character, e.g. P(O-*o*-tolyl)$_3$, give highest selectivities.

The use of sterically bulky ligands is an effective way to enhance dissociative substitution reactions. For example, the ligand dissociation constant K_1 for the reaction in Eqn 5.8 increases by a factor of

$$NiL_4 \overset{K_1}{\Longleftrightarrow} NiL_3 + L \qquad (5.8)$$

10^8 (from 6×10^{-10} to 4×10^{-2}) on replacing L = P(O-*p*-tolyl)$_3$ by P(O-*o*-tolyl)$_3$, with a corresponding increase in the ligand cone angle (θ) from only 128 to 141°; there are also minor electronic effects associated with this equilibrium constant. As a consequence, in the case of L = P(O-*o*-tolyl)$_3$, the fourth ligand is only very weakly held, leading to relatively high concentrations of the reactive 16e complex in solution, from which it has proved possible to both isolate and characterize the discrete coordinatively unsaturated complexes NiL_3 and $Ni(C_2H_4)L_2$.

Ligand dissociation from NiL_4 to NiL_3 is highly favoured in the case of P(O-*o*-tolyl)$_3$.

First stage

Mechanistically, the addition of the first molecule of HCN to buta-1,3-diene can be considered to occur in terms of the sequence of reactions outlined in Fig. 5.1, incorporating (*cf.* Section **2.1**):

- reversible ligand dissociation from NiL_4 to the coordinatively unsaturated 16e complex NiL_3,
- generation of an Ni(II) hydride by oxidative addition of HCN to the zerovalent nickel complex NiL_3,
- coordination of buta-1,3-diene,
- formation of a π-allyl intermediate through a *cis* rearrangement of the nickel hydride,
- reductive elimination of the two isomeric C_5 mononitriles, and
- regeneration of the catalyst by addition of further HCN.

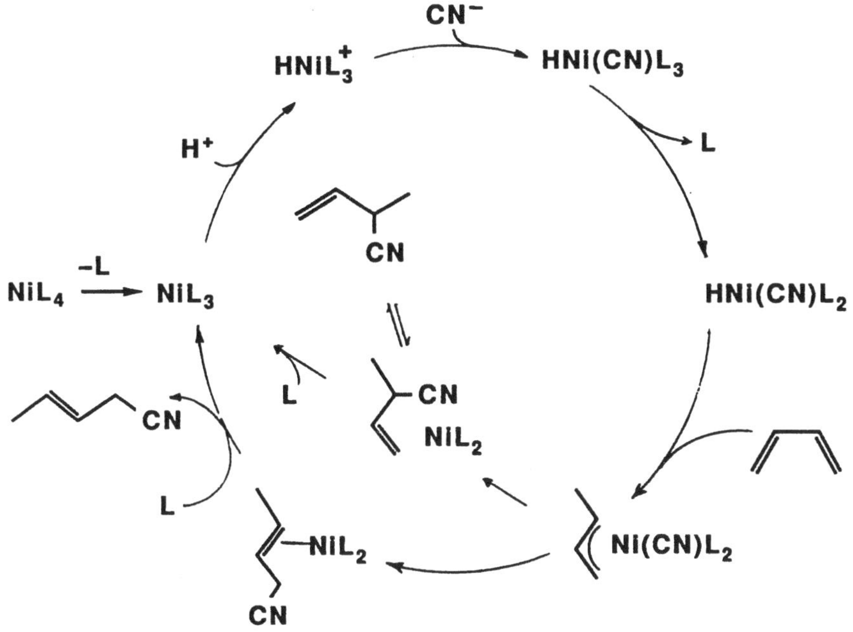

Fig. 5.1 The reaction mechanism of Ni-catalysed addition of HCN to buta-1,3-diene.

Supporting evidence for this scheme has been derived from the observation, by NMR spectroscopy at room temperature, of buta-1,3-diene complexation and insertion into the Ni–H bond, giving π-allylnickel cyanide species, which subsequently reductively eliminate the mixed C_5 mononitriles. Spectroscopic studies in which ethylene is substituted for buta-1,3-diene have provided further support for these reaction schemes by the detection of the discrete complexes

$C_2H_5NiL(C_2H_4)CN$, $(C_2H_4)NiL_2$, $(C_2H_4)NiL_3$, NiL_3, NiL_4 and $HNiL_3CN$, where L = P(O-*o*-tolyl)$_3$.

Second stage

The key role of Lewis acids in accelerating the isomerization of 2-methyl-3-butenenitrile to 3-pentenenitrile has been interpreted in terms of their ability to enhance the concentration of the cationic Ni hydride species which acts as the isomerization catalyst. NMR spectroscopic studies have confirmed that the Lewis acid co-catalysts (A) coordinate strongly to the nitrogen lone electron pair of the hydrido-cyanide intermediates, with the formation of adducts of the type $HNiL_3CN{\bullet}A$. Lewis acid-assisted removal of cyanide generates the cationic nickel hydride $[HNiL_3]^+$ (Eqn 5.9).

The rate of isomerization is enhanced by Lewis acid co-catalysts which increase the effective concentration of cationic Ni hydride $[HNiL_3]^+$ in the system.

$$HNi(CN{\bullet}A)L_3 \Longleftrightarrow [HNiL_3]^+[CN{\bullet}A]^- \qquad (5.9)$$

Thus, the rate of isomerization of 2-methyl-3-butenenitrile to the more highly thermodynamically favoured (by a factor of 9) 3-pentenenitrile is enhanced. The mechanism of isomerization of 2-methyl-3-butenenitrile is believed to occur via a relatively rare C–C cleavage reaction.

Third stage

This involves a combination of isomerization of 3-pentenenitrile to 4-pentenenitrile and migratory insertion of the more highly activated C=C bond in 4-pentenenitrile into the Ni–H bond, followed by reductive elimination of adiponitrile from the resulting π-allylnickel cyanide complex. The Lewis acid binds to the Ni–CN moiety and in this case the size of the Lewis acid has a direct bearing on product selectivity – bulkier Lewis acids favour the production of linear nitriles (*cf.* $ZnCl_2$ vs. BPh_3). The exact nature of the complex from which final reductive elimination occurs is still subject to speculation. However, based on model chemistry, it seems most probable that a five-coordinate Ni hydride of the type $L_2NiH(CN{\bullet}A)(R)$ is involved.

The rate of isomerization is again enhanced by the Lewis acid. In this case however steric effects assume additional importance in influencing the selectivity to linear products; BPh_3 is among the most effective.

In general, it is believed that the *net* effect of the Lewis acids on hydrocyanation is to increase the effective concentration of Ni in the catalytic loop, thereby accelerating the rate of C–C coupling, and the rate of formation of alkenenitriles from π-allylnickel cyanide complexes. An additional effect, contributed by bulky Lewis acids such as BPh_3, complements the steric effect of the bulky phosphite ligands in directing the second HCN addition in favour of the desired linear rather than branched products. This is presumably a consequence of destabilization of bulky branched alkyl intermediates relative to less crowded linear ones. In this way the addition of HCN to the non-conjugated double bond in an *anti*-Markovnikov manner is facilitated.

Impact on organometallic chemistry and homogeneous catalysis

It is important to acknowledge the significance of important fundamental concepts, now in everyday use by the practitioners of organometallic chemistry and homogeneous catalysis (Section **2.1**), that have emerged directly from fundamental supporting background research closely related to the development of this Ni-catalysed hydrocyanation process. These include the following:

- a realization of the significance of 16 and 18 electron intermediates in the operation of many catalytic processes,
- the elucidation and quantification of important electronic (χ) and steric (ligand cone angle, θ) parameters that influence ligand coordination and dissociation,
- the role of metal hydride addition/elimination mechanisms in olefin isomerization, and
- the use of Tolman cycles to describe catalytic sequences.

Several important concepts relevant to organometallic chemistry and homogeneous catalysis have emerged from fundamental research directly related to the development of this hydrocyanation process.

5.4 Process options for Nylon manufacture

Finally, to return to the present situation concerning the production of intermediates for the manufacture of Nylon, from the foregoing commentary it will be evident that the production of adiponitrile, via either propylene or buta-1,3-diene as feedstock can represent economically viable alternative options to the multistep routes based on benzene. Adiponitrile may be either hydrogenated to HMD or hydrolysed to adipic acid to provide both intermediates for the manufacture of Nylon.

The development of new process routes to adiponitrile has expanded the range of feedstock options available for the manufacture of Nylon since it can provide an intermediate for the manufacture of both adipic acid and HMD.

As in all multistage chemical processes, the economics depend on the balance of a number of factors including, in this case, the relative (and fluctuating) world prices of benzene, buta-1,3-diene, and propylene. In addition, geographical considerations, such as those countries where cheap electricity is available (*not* the UK and the majority of Europe!), can provide a significant commercial advantage for the electrohydrodimerization of acrylonitrile. In general, however, cyclohexane remains the preferred intermediate for adipic acid manufacture, but the position is less clear for adiponitrile/HMD where all three alternative feedstocks can provide commercially viable process options.

6 Olefin oligomerization and polymerization

6.1 Background

Olefin polymerization provides the principal industrial application of organometallic catalysis, with *ca.* 60% of production capacity accounted for by heterogeneous Ziegler–Natta catalysts.

Important applications of homogeneous catalysts include the production of α-olefins and polyketones, using Ni and Pd catalysts respectively.

The main industrial use for organometallic catalysts – but *not necessarily* homogeneous catalysts – is the polymerization of olefins. About 60 million tonnes of polyolefins are produced annually, 60% of this total with *heterogeneous* Ziegler–Natta catalysts and their derivatives. A considerably smaller, but none the less significant (1 mte y^{-1}), application which requires the use of homogeneous catalysts is the the Shell Higher Olefins Process (SHOP), which now accounts for *ca.* 50% of all α-olefins production. The key stage in this is the homogeneous Ni-catalysed oligomerization of ethylene to linear α-olefins. Polyketones, which can be produced by the co-polymerization of olefins with carbon monoxide, are of particular interest as speciality polymers, in the context of their superior materials performance properties relative to more conventional polyolefins. A new and highly selective process which utilizes soluble Pd catalyst precursors for the manufacture of alternating ethylene/CO co-polymers has recently been commercialized, also by the Shell group. The two processes are described in this chapter together with some exciting new advances in what was, until recently, believed to be the relatively mature field of olefin polymerization. These include two types of high-activity polymerization catalysts, capable of tunable regio- and stereo-selectivities, based on (i) metallocene complexes of the *early* transition metals and (ii) complexes of the *later* transition metals with multidentate N– and N$\frown$O ligands containing sterically bulky substituents; both are close to achieving significant commercial status.

Hitherto in this primer, almost all the homogeneously catalysed transformations that have been described involve reactions in which the addition of reactants, for example, CO and H_2, to a substrate, such as propylene, result in its *selective* conversion into a desired product, in one of two isomeric forms, e.g. *n*- and *iso*-butyraldehyde (Chapter **3**). In commercial operation this individual step may then be integrated into a more complex reaction sequence as part of a total chemical process, e.g. aldol condensation, dehydration, and hydrogenation, as required in the production of 2-ethylhexanol. With transformations such as dimerization, oligomerization, and polymerization, however, the term *selectivity* takes on a rather different meaning, because these classes of reaction comprise growth processes in which it is not generally possible to apply the principles of catalyst tuning, described previously, to provide a specific chain length according to particular requirements. Thus, the olefin growth reactions involved in oligomerization and polymerization are governed by a sequence of initiation, propagation, and termination steps (Eqn 6.1), and a *distribution* of oligomer/polymer chain lengths is produced. Only in special circumstances, such as dimerization and the formation of very high molecular weight polymers, can high selectivities be achieved.

Oligomerization and polymerization reactions are growth processes, which are characterized by a sequence of initiation, propagation, and termination steps. Similar selectivities to those typical of many homogeneously catalysed reactions are not generally realized.

$$\begin{array}{ccccccc} \longrightarrow & M(\mathbf{R})_{n-1} & \xrightarrow{k_1} & M(\mathbf{R})_n & \xrightarrow{k_1} & M(\mathbf{R})_{n+1} & \xrightarrow{k_1} \\ & \downarrow k_2 & & \downarrow k_2 & & \downarrow k_2 & \\ & \mathbf{R}_{n-1} & & \mathbf{R}_n & & \mathbf{R}_{n+1} & \end{array} \qquad (6.1)$$

$\mathbf{R}$ = monomer

The distribution of chain lengths for the growth reaction can be derived mathematically from the Schulz–Flory distribution according to Eqn 6.2

$$W_n = n\alpha^{n-1}(1-\alpha)^2, \quad \text{where } \alpha = k_1/(k_1 + k_2) \qquad (6.2)$$

where W_n represents the weight fraction, and k_1, k_2 are the rate constants for the propagation and termination steps respectively. A plot of $\ln W_n/n$ vs. C_n is linear, with a slope of $\ln\alpha$. Thus the k_1/k_2 ratio, which is of diagnostic value, can be obtained.

Qualitatively, three extreme situations may be recognized:

- $k_1 >> k_2$: – propagation dominates over termination and high molecular weight polymer is the result.
- $k_1 << k_2$: – termination is much more rapid than propagation and selective dimerization can be achieved.
- $k_1 \approx k_2$: – representative of the more typical intermediate situation, the rate constants are of comparable value, and one in which the Schulz–Flory distribution of C_n products is obtained.

With the possible exception of reaction and confinement of product within, e.g. the cavities of zeolites, the majority of attempts to overcome the intrinsic limitation in product selectivity have been unsuccessful.

Many attempts have been made to overcome this limitation in product selectivity, with little success. The first process for the industrial oligomerization of ethylene was developed by Ziegler in the early 1950s and comprised a chain growth reaction catalysed by Al alkyls. Reaction conditions were severe, typically 130–250 bar pressure and 200–250°C. The predominance of β-hydrogen elimination (Section **2.1**) gave linear olefins with $C_n = 4–8$ as the major products. Although a number of commercial olefin oligomerization processes involve alkyl aluminium growth reactions, by using a combination of homogeneous Ni-catalysed olefin oligomerization in tandem with other catalytic reactions, subsequent developments by Shell have been more successful in circumventing the limitations imposed by the Schulz–Flory product distribution.

6.2 Olefin oligomerization – Shell higher olefins process

This provides a very elegant example of how it has proved possible to work within the limitations of a polymerization reaction but still optimize on the higher value products by using a combination of oligomerization, isomerization, and metathesis stages. In this case the highest value products are detergent-range alcohols in the C_n range 12–15. SHOP was commercialized in 1977 when a 115,000 te y^{-1} capacity plant came into operation at Giesmar, Louisiana, and a second plant of 117,000 te y^{-1} capacity was brought on stream at Stanlow, UK, in 1981. Subsequent expansions, and extension of the technology, have now led to a current total capacity of *ca.* 1.0 mte y^{-1}. For example, application of the metathesis technology inherent in SHOP was extended in 1986, under the acronym FEAST (Further Exploitation of Advanced Shell Technology), to the production of monomers and intermediates for a range of applications in speciality chemicals manufacture. Thus, co-metathesis of cyclic dienes (e.g. cycloocta-1,5-diene) and ethylene produces high-value linear α,ω-dienes such as 1,9-decadiene.

SHOP comprises a three-stage process involving ethylene oligomerization, followed by isomerization and metathesis, for the production of detergent-range C_{11}–C_{14} olefins.

The Shell Higher Olefins Process operates in three steps, namely a homogeneously catalysed ethylene growth reaction, followed by isomerization and metathesis stages, both of which use heterogeneous catalysts. The products are optimized to give olefins in the $C_{11}–C_{14}$ range which are subsequently converted into detergent-range alcohols by hydroformylation (Chapter **3**). Not only is the chemistry of the homogeneous catalyst system of considerable interest in the context of organometallic chemistry but the integration of the separate stages into the overall process also illustrates some important principles.

Ethylene oligomerization

In the first step of the process ethylene is selectively oligomerized (Eqn 6.3) to give, in a geometric Schulz–Flory distribution, *even* carbon number *linear* (>99%) α-olefins (93–99%) using Ni-based homogeneous catalysts obtained from the reaction of *bis*(cyclooctadienyl) nickel with chelating P⌒O ligands of the general type $R_2PCH_2CO_2^-$ (R = phenyl or cyclohexyl), and o-$R'R''PC_6H_4CO_2^-$ (R′, R″ = alkyl and/or phenyl).

Chelating P⌒O ligands are essential components of the homogeneous Ni catalysts for high activity and selectivity to α-olefins.

$$C_2H_4 \xrightarrow[80-120^\circ C]{70-140\ \text{bar}} \textit{even}\ C_n\ \text{linear olefins}\ (n \geq 4\text{-}20+) \quad (6.3)$$

This combination of phosphorus- *and* oxygen-chelating functionalities, which is essential to the successful operation of the catalyst, gives high activities – values of the order of 0.6 mol ethylene mol^{-1} Ni s^{-1} have been quoted, enabling low metal concentrations (0.001–0.005 mol %) to be used. It also ensures solubility in the polar solvent (1,4-butanediol), which is largely immiscible with the α-olefin product, thus allowing facile separation from catalyst and reactants (Section **1.2**). An additional advantage of this method of operation is that the formation of branched olefins by secondary reactions, e.g. isomerization, is minimal because of the very low concentrations of α-olefins in the catalyst/solvent phase.

The immiscibility of the ethylene oligomer product with the catalyst-containing phase allows easy and continuous product separation and is believed to be the key contributary factor to successful commercial operation.

The reacting mixture thus comprises *three* phases, namely, solvent containing the catalyst, oligomer product, and ethylene gas. As hydrocarbon product is formed it separates from the catalyst-containing solvent phase as a separate layer. After reaction, ethylene is stripped, and the oligomer product phase is readily separated from the catalyst/solvent phase, which is in turn recycled directly to the oligomerization reactor.

The process represents the first industrial application of a biphasic liquid/liquid system in catalysis.

High partial pressures of ethylene are required for good rates of reaction and high linearity of the α-olefins product. The latter is characterized by a geometric growth factor K (*cf.* k_1/k_2 in Eqn 6.1), which is normally independent of alkyl chain length, and defined by Eqn 6.4.

$$K = \frac{\text{moles } C_{n+2} \text{ olefin}}{\text{moles } C_n \text{ olefin}} \quad (6.4)$$

Control of K is the key to success of the process since it determines both the composition of the α-olefin product and the average carbon number of the overall process (see Fig. 6.1). K may be maintained within the range 0.75–0.80 by variation in the nature of the catalyst. A value of 0.80 represents an 80% probability of adding C_2H_4 with consequent chain propagation, and 20% probability of elimination/termination with regeneration of the catalyst.

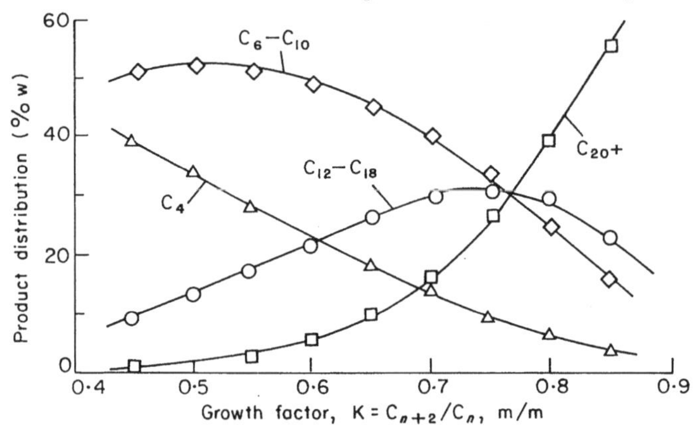

Fig. 6.1 Oligomer product distribution vs. *K* factor.

The pure C_{10}–C_{18} oligomer fraction is separated for sale as high-value α-olefins of high purity.

α-Olefin products are formed in the carbon number range n = 4-20+, many of which are outside the detergent range and of low commercial value. The C_{10}–C_{18} fraction, which in fact forms less than 30% of the total products, is separated for sale as high-purity α-olefins. The remainder, C_{4-8} and C_{20+} fractions, are then fed to the second stage.

Olefin isomerization

The second stage of the process comprises an isomerization step in which butene, C_6, C_8, and C_{20+} olefins are converted into a near-equilibrium distribution of straight-chain internal olefins over a typical heterogeneous isomerization catalyst (such as Na/K supported on alumina), under the reaction conditions indicated in Eqn 6.5.

α-Olefins of low commercial value are isomerized to a mixture of internal olefins in near-equilibrium distribution using a conventional heterogeneous catalyst.

$$RCH{=}CH_2 \xrightarrow[80-140^{\circ}C]{3.5-17\ \text{bar}} R'CH{=}CHR'' \ \text{(all possible isomers)} \qquad (6.5)$$

The reaction is believed to occur by a stepwise mechanism, and the double bond is shifted from terminal into internal positions approaching a near-equilibrium distribution. This is characterized by a *low* concentration of the double bond in the terminal (α) position and an almost statistical distribution over the other positions of the molecule. Thus a mixture of *internal* olefins in two discrete carbon number fractions, namely C_4–C_8 (larger fraction) and C_{20+}, is produced, which is then fed to the third stage of the process.

Olefin metathesis

This stage, which also involves a heterogeneous catalyst (Co-promoted MoO_3 supported on alumina), is the metathesis, or disproportionation, step in which a statistical redistribution of carbon frameworks occurs about the double bonds, e.g. Eqn 6.6, for the production of a C_{12} *internal* olefin.

Although homogeneous catalysts for olefin metathesis are well known, in this metathesis step a heterogeneous catalyst is preferred.

$$CH_3(CH_2)_8CH{=}CH(CH_2)_8CH_3 + CH_3CH{=}CHCH_3 \underset{80\text{–}140^\circ C}{\overset{13\ bar}{\Longleftrightarrow}} 2CH_3(CH_2)_8CH{=}CHCH_3 \quad (6.6)$$

A predominance of internal olefins is preferred in this stage because terminal (α) olefins reduce the catalytic efficiency through non-productive degenerate metathesis. This process yields about 10–15 wt % per pass of *odd*- and *even*-numbered linear *internal* olefins within the desired detergent-range carbon number; these products may then be separated by distillation as C_{11}–C_{14} linear olefins with a purity of >99% for both C_{11}/C_{12} and C_{13}/C_{14} fractions.

Since the metathesis stage only yields 10–15 wt% of the desired detergent-range product, extensive recycle of the lower (< C_{10}) and higher (> C_{15}) carbon number materials is required. After separation by distillation, the < C_{10} product is recycled to metathesis and the > C_{15} product returned through both isomerization *and* metathesis stages.

Internal olefins in the C_{11}-C_{14} range are separated by distillation for sale, and those of C_n < C_{10} and > C_{15} are recycled.

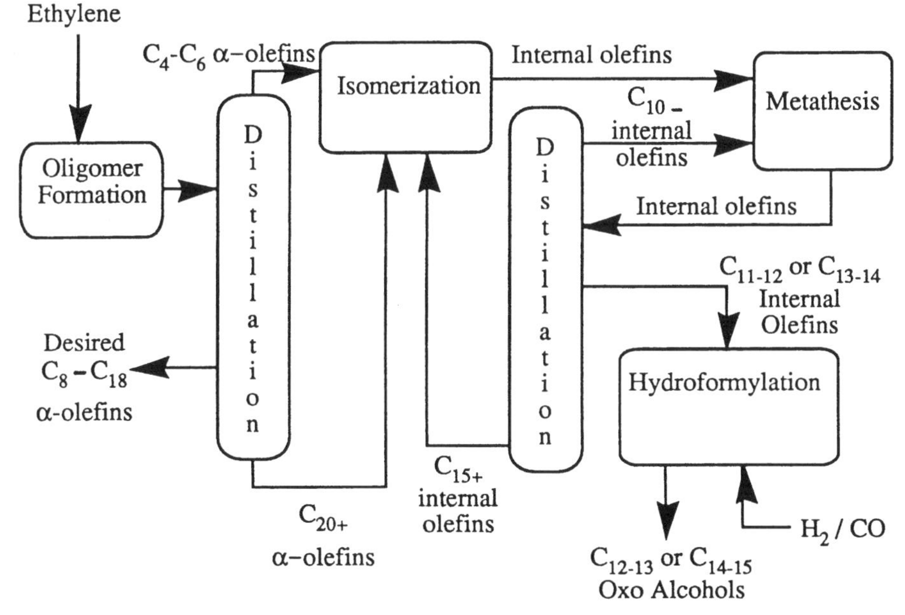

Fig. 6.2 Flow diagram of the Shell Higher Olefins Process. (Reprinted with permission from Wiley-Interscience).

The product olefins are subsequently converted into detergent alcohols by hydroformylation and hydrogenation. A simplified flow diagram of the process is presented in Fig. 6.2 to illustrate how integration of the various steps is accomplished.

Mechanism of Ni-catalysed oligomerization

The reaction mechanism of Ni-catalysed ethylene oligomerization incorporates several of the concepts outlined in Section **2.1**.

It is widely accepted that the active catalyst in ethylene oligomerization is a Ni–H species, generated in situ under reaction conditions as indicated in Fig. 6.3. Supporting evidence for the existence of a nickel hydride has been obtained from in situ NMR spectroscopic studies in some cases. Formal insertion of ethylene into the Ni–H bond yields a Ni–ethyl complex which can either undergo additional multiple insertions of ethylene, and hence propagate the oligomerization, or the intermediate Ni-alkyls may β-eliminate to release the corresponding α-olefin. The specific importance of the chelating P$^\frown$O functionality appears to be related, at least in part, to controlling the selectivity of the reaction, the additional ligands simply serving to stabilize the complexes.

p = propagation
e = elimination

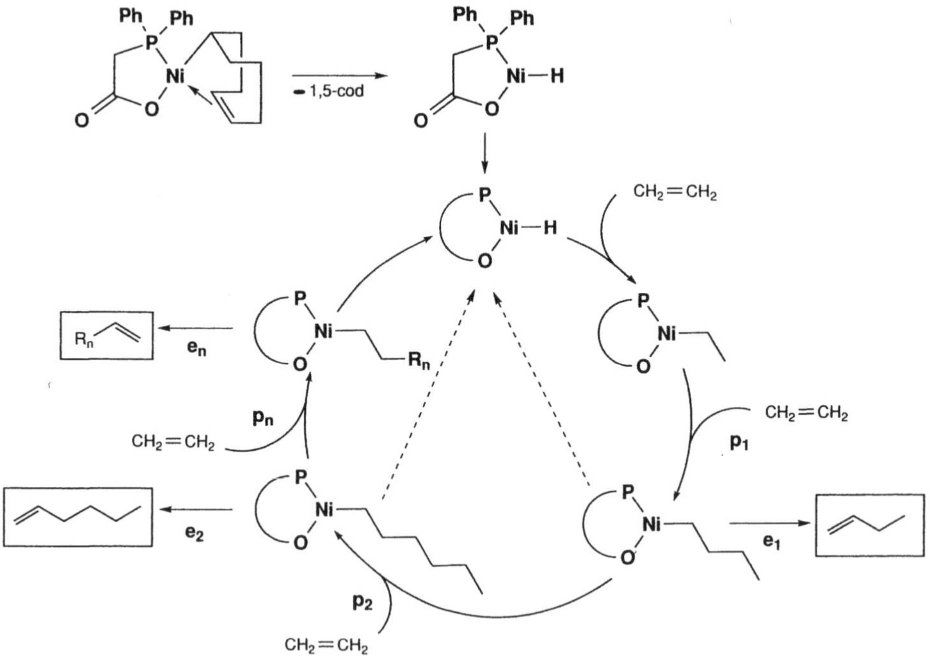

Fig. 6.3 The proposed mechanism of Ni-catalysed ethylene oligomerization. (Reprinted with permission from Wiley-VCH).

In summary, SHOP represents an elegant example of how it is possible to work within the limitations of the Schulz–Flory product distribution of an oligomerization/polymerization-type reaction, yet still optimize on the higher value products by using the sequence of reactions described. The process thus solves the problem of the basic mismatch between product distribution from an oligomerization process and market requirements. Today SHOP accounts for about 50% of the linear α-olefins produced by oligomerization and thus constitutes a very significant application of homogeneous catalysis.

6.3 Polyketone production

The co-polymerization of C_2H_4 and CO, both readily available feedstocks, to give the 1:1 perfectly alternating co-polymer (poly-3-oxotrimethylene, **6.1**) is of interest because of the useful properties exhibited by such polyketones in thermoplastic materials applications.

$$—[CH_2CH_2C(O)]_n— \qquad \textbf{6.1}$$

These include strong rigidity and impact strength, good chemical resistance to acids, bases, and solvents, as well as a high tracking resistance, and stability against electrolytic corrosion. In these respects polyketones are superior to polyolefins, polyamides, and polyacetals. The industrial production of polyketone by Shell under the trade-name 'Carilon', in 1996, represents one of the most recent commercial applications of homogeneous catalysis.

Polyketones are melt-processable and provide a new range of desirable engineering thermoplastic polymers.

The formation of non-alternating polyketones by the co-polymerization of C_2H_4 and CO has been known since the late 1940s when these polymers were produced under extreme pressures (500–1500 bar) using free radical initiation. The polymers were generally of low molecular weight, with branched molecular structures, and irregular CO incorporation. As a consequence, they had poor physical properties and were highly soluble in common organic solvents. The first metal-catalysed co-polymerization of C_2H_4 and CO, to give alternating ethylene–carbon monoxide structures, was described by Reppe in 1951, and used $K_2Ni(CN)_4$ in water at 100–200°C. Low melting oligomers, together with diethyl ketone and propionic acid, were produced. The first Pd-catalysed alternating co-polymerization was discovered by ICI, but required severe reaction conditions, and gave low yields of polyketone.

Earliest routes to polyketones gave low molecular weight non-alternating co-polymers with poor physical properties, e.g. low m.p., and high solubility in organic solvents.

During studies of the reactivity of cationic Pd(II) complexes containing tertiary phosphine ligands and Bronsted acids of weakly, or non-coordinating anions, e.g. *p*-tosylate, in methanol, as catalysts for the methoxycarbonylation (Section **3.5**) of ethylene to methyl propionate (Eqn 6.7), it was found that substitution of *monodentate*

$$CH_2{=}CH_2 + CO + CH_3OH \xrightarrow{[Pd]} CH_3CH_2CO_2CH_3 \qquad (6.7)$$

tertiary phosphines by *bidentate* analogues, e.g. 1,3-*bis*(diphenylphosphino)propane (dppp), unexpectedly gave no methyl propionate; instead, a high molecular weight (M_n *ca.* 20,000) perfectly alternating 1:1 ethylene/CO copolymer was formed in 100% selectivity at very high reaction rates (*ca.* 6000 g per g Pd h^{-1}), Eqn 6.8.

This process for the production of an ethylene/CO co-polymer constitutes an example of a transition metal catalysed polymerization with almost perfect control over selectivity (*cf.* Section **6.1**).

$$n(CH_2{=}CH_2) + n(CO) + CH_3OH \xrightarrow{[Pd]} H{-}[CH_2CH_2CO]_n{-}OCH_3 \quad (6.8)$$

Thus, exchange of *mono*- for *bi*-dentate phosphine ligand has resulted in an unprecedented switch in reaction chemistry from simple methoxycarbonylation to co-polymerization and polyketone formation. The product is high melting (*ca.* 260°C) and insoluble in most organic solvents, crystallizing and precipitating during co-polymerization. The Pd catalysts are extremely active, giving typical yields of 10^4 mol. of ethylene converted per mol. Pd h^{-1} under relatively mild, economically viable, reaction conditions (45 bar, 90°C).

Under optimum conditions, conversions of greater than 10^6 mol. ethylene per mol. of Pd can be obtained.

Variation in the nature of the bidentate phosphine ligand L_2 provides significant variations in both the reaction rate and molecular weight of the product. Counterions also affect reaction rates, the highest activities being obtained with weakly or non-coordinating anions such as tosylate or triflate. The catalysts are simple to prepare, either separately or in situ, by reaction of $L_2Pd(OAc)_2$ with HX (X = *p*-tosylate, triflate, sulphonate, etc.) in methanol, the solvent of choice for co-polymerization. This largely unprecedented discovery of the combined importance, for high catalytic activity, of bidentate ligands and weakly coordinating anions around a cationic Pd(II) centre has thus provided, for the first time, access to a commercially viable route for the production of polyketones.

Reaction mechanism

Details of co-polymerization and, in particular, the cross-over between co-polymerization and methoxycarbonylation, pose some intriguing mechanistic questions. A conceptually reasonable picture of the key steps including initiation, propagation, and termination, together with a basic understanding of the respective roles of the bidentate ligands, and anions, in controlling the alternation between CO and ethylene insertion, has been proposed. In considering mechanistic aspects, it is important to emphasize the close relationship, which is not perhaps immediately obvious, between methyl propionate formation and co-polymer production. The principal difference between the two is simply the inclusion of a propagation step in the latter mechanistic pathway, the course of the reaction being determined by the competition between propagation and termination (*cf.* Section **6.1**).

Polyketone formation can be regarded in terms of methyl propionate formation with the inclusion of a propagation step. The preferred route thus depends on the competition between propagation and termination (Section **6.1**)

Initiation

End-group analyses of the co-polymers, by ^{13}C NMR spectroscopy, have shown the presence of 50% ester ($-COOCH_3$) and 50% ketone ($-COCH_2CH_3$) groups, in accordance with the average product composition. However, analysis of co-formed oligomer by-products obtained with some catalysts has revealed the formation of both diester and diketone products, i.e. crossover products, in addition to the simple keto-ester. This requires the operation of two initiation and two termination mechanisms for polyketone formation. These are most simply explained in terms of the chemistry of methoxycarbonylation, in which the operation of *both* methoxycarbonyl- and hydride-based catalytic cycles has been proposed (Section **3.5**). *Two* equivalent initiation pathways seem viable, namely via Pd-methoxycarbonyl or Pd-hydride species. The first produces ester end-groups and starts from a Pd-methoxycarbonyl species which can be formed either by CO insertion into a Pd methoxide or by direct attack of methanol on coordinated CO (Chapter **3**, Fig. 3.2). Alternatively, a polymer chain can start by 'insertion' of ethylene into a Pd hydride, ultimately producing a ketone end-group. Migratory insertions of ethylene into the Pd hydride and of CO into the resulting ethyl complex are both rapid and *reversible*; it is believed that the second ethylene insertion (into the Pd acyl) is irreversible and 'traps' the acyl to start the chain (Eqn 6.9).

Detailed co-polymer end-group analyses have led to the conclusion that two separate initiation and termination mechanisms are necessary to account for the product composition.

Initiation appears to involve both methoxycarbonyl and hydride mechanistic pathways which generate ester and ketone end-groups, respectively, in the product polyketone.

$$[L_2PdH]^+ \overset{C_2H_4}{\Longleftrightarrow} [L_2Pd(CH_2CH_3)]^+ \overset{CO}{\Longleftrightarrow} [L_2Pd(COCH_2CH_3)]^+ \overset{C_2H_4}{\longrightarrow}$$

$$[L_2Pd(CH_2CH_2COCH_2CH_3)]^+ \longrightarrow\longrightarrow \quad (6.9)$$

Propagation

The catalytically active species in polyketone formation is thought to be a d^8 square planar cationic Pd complex $[L_2Pd(X)\mathbf{P}\sim]^+$, where L_2 represents the *cis*-chelating bidentate ligand and $\mathbf{P}\sim$ the growing polymer chain. The fourth coordination site (X) at Pd may be occupied by an anion, solvent molecule, a carbonyl group of the chain, or a monomer molecule. The two alternating propagation steps are migratory insertion of CO into the Pd–alkyl bond (Eqn 6.10), followed by migratory insertion of C_2H_4 into the resulting Pd–acyl bond (Eqn 6.11).

$$[L_2PdCH_2CH_2\mathbf{P}\sim]^+ + CO \longrightarrow [L_2PdC(O)CH_2CH_2\mathbf{P}\sim]^+ \quad (6.10)$$

$$[L_2PdC(O)CH_2CH_2\mathbf{P}\sim]^+ + CH_2{=}CH_2 \longrightarrow$$

$$[L_2PdCH_2CH_2C(O)CH_2CH_2\mathbf{P}\sim]^+ \quad (6.11)$$

Propagation requires sequential migratory insertions of CO and ethylene into the growing polymer chain.

A particularly remarkable feature of the chemistry is the apparent lack of propagation errors such as double CO or ethylene insertion reactions. Migratory insertion of CO into metal–carbon bonds is a mechanistic feature which is common to many homogeneously catalysed reactions (Chapters **2–4**) and, under polymerization conditions, CO insertion is thought to be rapid and reversible. Olefin insertion into a Pd–C bond, although known, is less common but does appear to be facile, particularly into Pd–acyl bonds. It seems likely, therefore, that ethylene insertion is the slowest and irreversible rate-determining step in polyketone formation.

Termination

Termination also involves two competing reactions, methanolysis of Pd–alkyl and Pd–acyl intermediates, giving ketone and ester end-groups in the polyketone.

As with the initiation of CO/ethylene co-polymerization, two termination pathways are required, and these both involve methanolysis, of Pd alkyls and acyls respectively. Methanolysis of a Pd–alkyl bond produces polymer containing a saturated ketone end-group (Eqn 6.12), with regeneration of the methoxycarbonyl initiator by insertion of CO into the Pd–OCH_3 bond.

$$[L_2PdCH_2CH_2CO\mathbf{P}\sim]^+ \xrightarrow{CH_3OH} [L_2PdOCH_3]^+ + CH_3CH_2CO\mathbf{P}\sim \quad (6.12)$$

Conversely, methanolysis of a Pd–acyl bond gives an ester end-group (Eqn 6.13), with regeneration of the Pd hydride which acts as initiator for the next polymer chain.

$$[L_2PdCO\mathbf{P}\sim]^+ \xrightarrow{CH_3OH} [L_2PdH]^+ + \sim\mathbf{P}COOCH_3 \quad (6.13)$$

The reaction chemistry and product distribution may be rationalized in terms of a combination of the hydride and methoxycarbonyl mechanistic cycles.

An overall reaction scheme encompassing the hydride (A) and methoxycarbonyl (B) cycles is summarized in Fig. 6.4. The hydride and methoxycarbonyl mechanistic cycles both give keto-ester molecules but the cycles are connected by two 'cross' termination steps which give diester and diketone products, consistent with the observed product distribution.

A final point concerns the role of oxidant promoters such as benzoquinone in these reactions. Their inclusion is found to enhance reaction rates by factors of 2–15 for the different diphosphine ligands. Although they have no effect on chain length (and therefore propagation and termination steps), they do result in a greater proportion of ester end-groups, suggesting that a greater proportion of growing chains start via methoxycarbonyl species, thus *possibly* favouring the methoxycarbonyl pathway. Nevertheless, knowledge of the product distribution alone is insufficient to establish which cycle dominates the chemistry and further mechanistic information is required.

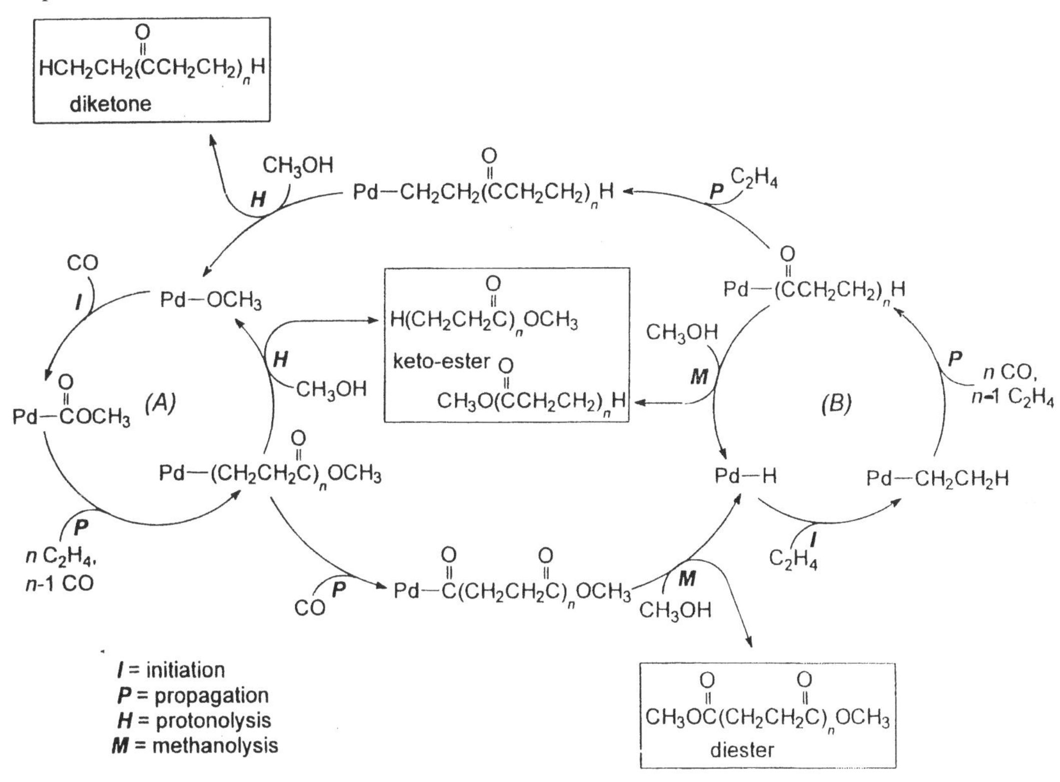

Fig. 6.4 Proposed mechanism of Pd-catalysed polyketone formation. (Reprinted with permission from the American Chemical Society).

Role of ligands and counter ions

The use of *cis* chelating bidentate phosphine ligands appears to be a prerequisite for the selective production of polyketone (although there is *at least* one exception to this generalization). This may be a consequence of the constraint, to a mutually *cis* orientation, of the growing polymer chain and 'vacant' fourth coordination site around Pd imposed by the bidentate ligand, which provides a preference for insertion over elimination. It also ensures that the growing polymer chain-end and incoming monomers will be in the *cis* configuration required for chain propagation. In contrast, with monodentate phosphine ligands, *cis–trans* isomerization is possible, and *trans* geometries are actually preferred on electronic grounds, thus allowing ready methanolysis of alkyl or acyl intermediates in preference to insertion of CO/ethylene. The weakly coordinating anions are also key components of the catalyst composition and their role has been

The use of *cis* chelating bidentate phosphine ligands ensures that the growing polymer chain and incoming monomers, CO and ethylene, are in the *cis* configuration required for chain propagation.

Significant concentrations of the thermodynamically more stable *trans* isomers, made accessible by the use of monodentate phosphine ligands, may promote ready methanolysis of the acyl intermediate (and the formation of methyl propionate) in preference to migratory insertion of CO/ethylene.

interpreted in terms of enhancement of the overall electrophilicity of the cationic Pd centre by ion-pairing effects, mediated to some degree by the polarity of the reaction solvent, which is usually methanol.

The explanation for the perfect 1:1 alternation by CO and ethylene in chain propagation is not entirely clear, particularly since, in the absence of CO, the same catalysts can show even higher activity for the dimerization of ethylene to butenes, as a consequence of which double insertion of ethylene might be considered to be highly probable. Nevertheless, this particular process for the production of an ethylene/CO co-polymer constitutes an example of transition metal-catalysed polymerization with almost perfect control over selectivity (Section **6.1**). These catalysts have also been demonstrated to catalyse the alternating *co*- or *ter*-polymerization of higher olefins, including propylene, styrene, and dicyclopentadiene, thus providing considerable scope for further development and refinement of this new range of engineering thermoplastic polymers.

This class of catalysts is also effective for the alternating *co*- and *ter*- polymerization of higher olefins.

6.4 Olefin polymerization

As indicated in Sections **1.3** and **6.1**, the main industrial use for organometallic derived catalysts is in the manufacture of high molecular weight polyolefins (i.e. the situation in which $k_1 >> k_2$ in Eqn 6.1). About 60 million tonnes of polyolefins are produced annually, using *heterogeneous* catalysts derived from organometallic compounds, and, following 40 years of development, the standard catalysts are highly efficient.

By far the largest-volume plastic is polyethylene (accounting for *ca.* 40 mte per annum), originally discovered in ICI in 1933 and obtained by free radical polymerization under severe reaction conditions (2000 bar, 200°C). The resultant polymer was highly branched, of low density (0.91–0.93), and melted over a wide temperature range. Currently two principal varieties are produced, by coordination, rather than free radical polymerization. These comprise high-density (0.97) polyethylene (HDPE), which is a linear polymer with a m.p. of about 136°C, and linear low-density polyethylene (LLDPE), which represents a family of polyethylenes intermediate between high- and low-density materials. The latter are prepared for specific applications by modification of the conditions used for the manufacture of high-density products.

In commercial terms the most significant and highest volume olefin polymers are polyethylene and polypropylene.

Other important, but lower volume, polymers obtained using organometallic-derived catalysts are polypropylene (*ca.* 20 mte y^{-1}), ethylene/propylene/diene elastomers, and polybutadienes (Table 6.1).

Polypropylene offers applications advantages over polyethylene although the highly desirable *stereospecific* polymerization of propylene represents a considerably more difficult target than the polymerization of ethylene. Suitable catalysts are therefore considerably fewer in number.

The elements traditionally used for commercial polymerization catalysts are those of the early transition metals, particularly Ti, V, Zr, and Cr.

Table 6.1 Polyolefin production with organometallic-derived catalysts

Polymer	*Major catalysts*
High-density polyethylene	$TiCl_x/AlR_3$, Cr/silica
Linear low-density polyethylene	$TiCl_x/AlR_3$, Cr/silica
Polypropylene	$TiCl_3/AlR_2Cl$, $TiCl_4/MgCl_2$
Ethylene/propylene/diene elastomers	$VOCl_3/AlR_2Cl$, $TiCl_4/MgCl_2/AlR_2Cl$, $ZrCp_2Cl_2$/MAO*
cis-1,4-Polybutadiene	TiI_4/AlR_3, $Co(O_2CR)_2/Al_2R_3Cl_3$, $Ni(O_2CR)_2/AlR_3/BF_3$

*MAO = methylaluminoxane, Cp = cyclopentadiene.

Ziegler–Natta catalysts

Classical Ziegler–Natta catalysts, for the discovery of which the Nobel prize was awarded to the inventors in 1963, are *heterogeneous* materials formed by interaction of $TiCl_3$ and Al alkyls, or $TiCl_4$ supported on $MgCl_2$, with $AlEt_3$ as co-catalyst, and account for about 60% of the worldwide production capacity of polyolefins. They are active at 25°C and 1 bar, in sharp contrast to the severe conditions required for the original thermal, radical-initiated, polymerization of ethylene. Such heterogeneous catalysts are complex systems, with different active sites, and the polymer structure can be influenced only to a limited degree. In addition, non-uniformity of the active sites in these heterogeneous catalysts renders mechanistic study and rational design of modified catalysts extremely difficult. Notwithstanding the fact that heterogeneous catalysts are preferred, largely because of chemical engineering requirements (Section **1.2**), among the great number of Ziegler–Natta catalysts, those that comprise *nominally* homogeneous systems have been preferentially studied in the past in order to attempt to understand the elementary steps of polymerization. Nevertheless, even these are not necessarily soluble under reaction conditions, and many questions remain unanswered concerning the mechanistic details of coordination polymerization. For example, basic

The vast majority of polymers are manufactured using heterogeneous two-component Ziegler–Natta catalysts prepared by the activation/reduction of early transition metal halides with aluminium alkyls.

Many questions remain unanswered concerning the intimate details of reaction mechanisms for the majority of the classical polymerization catalysts.

facts such as the oxidation state and structure of the catalytically active metal site, and the nature of the insertion mechanism, are not well established in many cases. Their mode and mechanism of action remain topics of current research.

Alternatives to Ziegler–Natta catalysts, specifically for the manufacture of polyethylene (they are inactive towards propylene), are the chromium-based catalysts, e.g. Cr/silica, developed by Phillips. These are less active than Ziegler–Natta catalysts and give high molecular weight rigid polymethylene-type polymers. One preparative route to the Phillips catalysts involves the anchoring of metallocenes such as *bis*(π-cyclopentadienyl)Cr to silica by interaction with the surface hydroxyl groups, and provides an illustration of one of the first industrial applications of the concept of surface organometallic chemistry (Section **2.5**) – and indeed pre-dated the terminology by at least two decades!

In view of these later developments it is important to recognize that conventional Ziegler-Natta catalysts require strictly anhydrous olefin feeds, since aluminium alkyls react extremely vigorously with traces of water.

Perhaps of more direct relevance to applications of *homogeneous* catalysis, the most recent developments in polymerization chemistry concern (i) the use of methylaluminoxanes (MAOs) as activators to provide a range of highly active metallocene catalysts and (ii) the unprecedented (in the light of prior knowledge of polymerization catalysts) discovery of highly active complexes of the *later* transition metals, particularly Fe, Co, and Ni, containing multidentate N– and N$^\frown$O ligands bearing bulky substituents. Some of the most recent do not even require activation by MAO and may therefore be described as truly 'single-site' catalysts. Both these developments are beginning to have an impact on commercially produced polymers.

New 'homogeneous' metallocene-based olefin polymerization catalysts

Although homogeneous ethylene polymerization catalysts based on *bis*(π-cyclopentadienyl) Ti(IV) compounds, using $AlEt_2Cl$ or $AlEt_3$ as co-catalysts, have been extensively studied, their activities are relatively poor (less than 200 g polyethylene (mol. catalyst)$^{-1}$ h^{-1}). Likewise, both homogeneous and alumina-supported Ti(benzyl)$_4$, M(π-allyl)$_4$, M = Zr and Hf, although much more active and long-lived in the solid state, have not achieved commercial status.

The advent of MAO co-catalysts has facilitated the generation of metallocene catalysts of 10–100 times higher activity than classic Ziegler–Natta systems.

This situation was transformed by the discovery, in 1977, of a new generation of so-called 'single-site' homogeneous catalysts, based on combinations of metallocenes, particularly derivatives of Cp_2ZrCl_2 (which have proved more active than their Ti or Hf analogues), activated with MAO co-catalysts, of approximate composition $[MeAlO-]_n$, formed by the controlled hydrolysis of $AlMe_3$. Both metallocene and MAO, as well as the active complex, are hydrocarbon soluble and the catalysts are up to 10^2 times more active than common heterogeneous counterparts. Thus, using Cp_2ZrCl_2 and MAO, polyethylene may be produced at rates of up to 40,000 kg g^{-1} Zr h^{-1} under mild reaction conditions (2.5 bar C_2H_4, 30°C, metallocene concentration 6.25 x 10^{-6} M in toluene, MAO/metallocene ratio 250:1). Substituted *bis*(indenyl) systems and bridged *bis*(fluorenyl)-zirconocenes in particular show very high activities, exceeding those of sterically less-hindered Cp_2ZrCl_2. The catalysts have extended lifetimes, retaining activity after up to 100 h on stream. Polyethylenes produced by such metallocene catalysts feature a molecular weight distribution of M_w/M_n = 2; 0.9–1.2 methyl groups per 1000 C atoms, 1.1–1.8 vinyl and *trans*-vinyl groups per 100 C atoms.

Using these catalysts it has proved possible to tailor the microstructure of the polymers (*cf.* Ziegler–Natta catalysts) by tuning the ligands. Also, molecular weight is easily modified, e.g. lowered, by increasing the temperature and ratio of metallocene:ethylene, or through the addition of small amounts of hydrogen (0.1–2 mol%).

In addition to the generation of higher catalytic activities it has proved possible to tailor the microstructure of the polymers by fine-tuning of the ligands.

Not only can polyethylene can be produced, but it is also possible to co-polymerize ethylene with α-olefins such as propylene, but-1-ene, pent-1-ene, hex-1-ene, and oct-1-ene, to produce LLDPE. The product parameters are consistent with the random incorporation of the co-monomer. In addition, many kinds of co-polymers and elastomers, new structures of polypropylenes, polymers and co-polymers of cyclic olefins can be obtained. In particular, catalysts of this type containing chiral centres are showing promise in the stereospecific polymerization of propylene to the desired isotactic product.

The first commercial applications (by Dow and Exxon) of the zirconocene/ MAO-derived catalysts include the production of elastomers by the ter-polymerization of ethylene, propylene, and dienes, and of polyethylenes with different properties from those polymers typically obtained using Ziegler-Natta catalysts.

Composition and role of methylaluminoxanes

The methylaluminoxanes are formed by the controlled reaction of $AlMe_3$ and water, with elimination of CH_4, and have the approximate composition $[MeAlO-]_n$, with a molecular mass in the range 1000–1500 g mol^{-1}. They contain linear, cyclic, and cross-linked compounds probably comprising predominantly four-coordinate Al centres and some $-OAlMe_2$ end-groups (**6.2**).

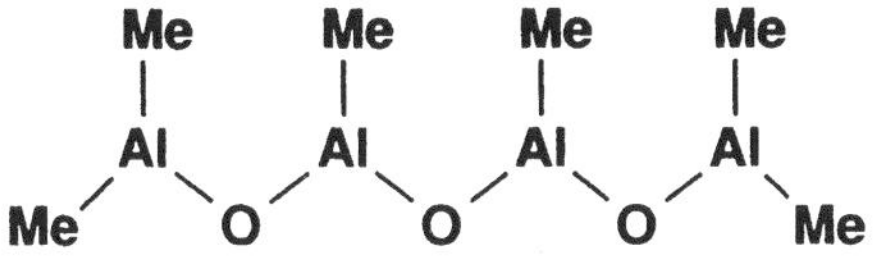

Me Me Me Me
Al Al Al Al
Me O O O Me

6.2

Excess of MAO is normally required, typical Al:metallocene ratios ranging between 50–100 for supported systems and 400–20,000 in solution. Among the different aluminoxaneco-catalysts, methylaluminoxane is much more effective than the ethyl- or di-*iso*-butyl analogues. The optimum Al/metallocene ratio depends on the metallocene used and

These systems should be considered as colloidal, rather than true homogeneous catalysts.

the experimental conditions. In reality, therefore, these systems are best viewed as *colloidal* rather than truly homogeneous catalysts. Indeed, in commercial use they are 'heterogenized' by supporting on silica or alumina pretreated with MAO.

The methylaluminoxane co-catalyst acts as an alkylating and CH_3^- abstraction agent, a Lewis acid to remove Cl^-, a bulky counterion, and as a scavenger to remove impurities.

The aluminoxane activators serve several functions, including alkylation of the metallocene component (Eqn 6.14), which occurs within seconds even at –60°C,

$$Cp_2MCl_2 \xrightarrow{MAO} Cp_2M(CH_3)Cl \xrightarrow{MAO} Cp_2M(CH_3)_2 \quad (6.14)$$

and formation of the active catalyst by abstraction of Me^- (Eqn 6.15).

$$Cp_2M(CH_3)_2 \Longleftrightarrow Cp_2M(CH_3)(CH_3Al_{MAO}) \Longleftrightarrow [Cp_2M(CH_3)]^+ + [(CH_3)Al_{MAO}]^- \quad (6.15)$$

For M = Ti, Zr, and Hf, the resulting catalytically active species is therefore a 14-electron cationic alkylmetallocenium ion formed by dissociation of the metallocene aluminoxane complex. The [aluminoxane–Me]$^-$ anion is considered to be weakly coordinating or even non-coordinating. The reaction mechanism is believed to involve successive additions and insertions of ethylene at each Zr centre comprising the 'single-site' catalyst. In summary, therefore, these highly active catalysts produce uniform homo- and co-polymers with narrow molecular weight distributions and polymer structure which may be controlled by the symmetry of the catalyst precursors.

Additional commercial applications of these metallocene-based catalysts are anticipated.

Later transition metal catalysts for olefin polymerization

A fundamental limitation of the Ziegler–Natta and metallocene-based catalysts is the highly electrophilic nature of the early transition metals which generally makes it impossible to use olefins containing polar functional groups as monomers or co-monomers. In addition, because of their highly oxophilic nature, a practical disadvantage is their susceptibility to deactivation by a range of poisons. These include traces of O_2, H_2O, and CO, which in turn requires that the monomer, solvents, and co-catalysts must be scrupulously purified before use.

The later, more typical, transition metals such as Fe, Co, and Ni are considerably less oxophilic, and therefore much more resistant to such deactivation processes. They also display much greater functional group tolerance, thus offering the possibility of polar co-monomer incorporation, e.g. acrylates. However, their ready susceptibility to chain termination by β-elimination (Section **2.1**) has hitherto largely precluded commercial applications in polymerization and restricted it to olefin oligomerization (Section **6.1**).

Later transition metals are less susceptible to poisoning than earlier transition metal polymerization catalysts, but susceptibility to β-elimination has hitherto restricted their applications to olefin oligomerization.

A recent, and largely unprecedented, advance is the demonstration that, with appropriate attention to the surrounding ligand environment, later transition metals can indeed provide extremely active catalysts for olefin polymerization. Thus, it has been demonstrated that the Fe complexes (**6.3**), and their Co, Ni, and Pd(II) analogues, containing tridentate bis(imino)pyridyl ligands of the type $\{[(2,6\text{-}ArN{=}CMe)_2C_5H_3N]MX_2\}$, where Ar = $2,6\text{-}C_6H_3(Pr^i)_2$ prepared by Schiff base condensation, are, after activation with MAO (at considerably lower MAO:M ratios than those required by the zirconocene-based systems) capable of polymerizing ethylene to high molecular weight polymers. The key to high polymer production using these aryl-substituted α-di-imine systems is believed to be the incorporation, on the aryl rings, of bulky *ortho* substituents which greatly retard the rate of chain transfer and β-elimination.

The key to the generation of polymerization rather than oligomerization catalysts is the use of bulky substituents on the ligands which have the effect of inhibiting termination by β-elimination.

The report of the use of Fe-based catalysts for the polymerization of ethylene for the first time is particularly notable. They are highly active and selective, giving products ranging from liquid α-olefins to high molecular weight polyethylene depending on the nature of the aryl substitution pattern.

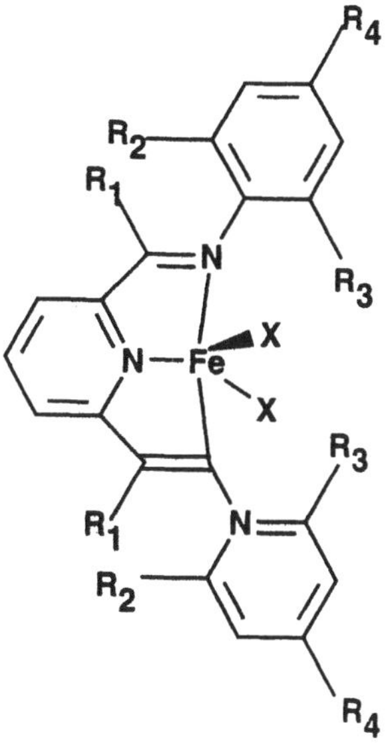

X = Cl and Br

6.3

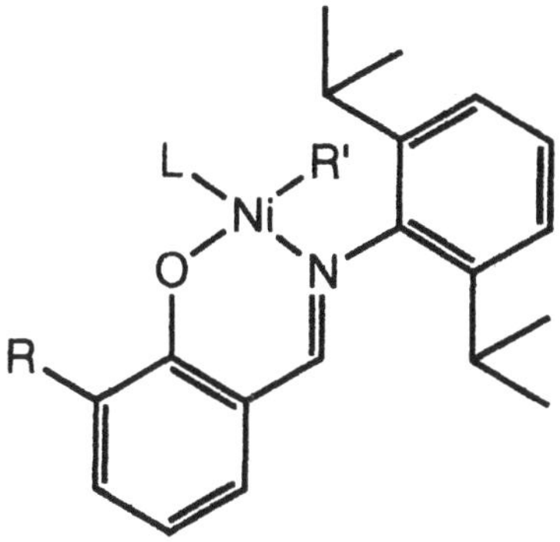

L = PPh_3, R = anthracene, R′ = Ph

6.4

The idea that late transition metal chelating complexes could polymerize α-olefins has stimulated a search for other ligand systems. Another significant development in this area concerns a report, in

The most recent development is the first example of a highly active, *truly* one component polymerization catalyst, i.e., one that does *not* require activation by MAO.

January 2000, of the design of a family of highly active late transition metal catalysts that are tolerant of both heteroatoms and reagents which have not been subject to stringent purification procedures. Furthermore, they are claimed *not* to require activation by MAO, and thus represent the first example of *truly* one-component polymerization catalysts. A typical ethylene polymerization catalyst precursor (**6.4**) contains a functionalized bidentate salicylaldimine N⌒O chelating ligand, with PPh_3 and phenyl groups as ancillary ligands coordinated to the Ni centre. The resulting polyethylene is of high molecular weight and exhibits a low degree of branching in the polymer chain. Most importantly, functionalized olefins may be polymerized. Typical catalytic activities of these one-component polymerization catalysts, at 7 bar C_2H_4 and ambient temperature, are compared with the MAO activated systems in Table 6.2.

Table 6.2 Comparison of typical activities observed with new ranges of ethylene polymerization catalysts

Catalyst precursor	*Co-catalyst*	*Activity in ethylene polymerization (kg polyethylene mol(metal)$^{-1}$ h^{-1})*
$[Cp_2ZrMe]^+[B(C_6H_5)_4]^-$	MAO	4500
$[(ArN{=}CMe)_2C_5H_3N]NiBr_2$	MAO	3000
6.4	none	3700

In conclusion, these new developments with catalysts containing the later transition metals have highlighted some attractive advantages over both the existing Ziegler–Natta and metallocene catalyst ranges. In addition to tolerance of functional groups on the substrate, they also offer considerable scope for tailoring/fine-tuning of activity and stereospecificity by controlled variation of functional groups attached to the salicylaldimine ligands. However, a specific issue that requires resolution concerns the poor thermal stability of these new catalysts in relation to the typical operating temperatures of 120°-140°C required by polyolefin manufacturers. Probably the most attractive commercial option at present is the use of iron-based catalysts for the oligomerization of ethylene to long chain α-olefins (*cf.* Section **6.2**).

7 Fine chemicals manufacture

7.1 Homogeneous catalysis and fine chemicals

As is evident in previous chapters, one of the great advantages to be derived from the use of homogeneous catalysts concerns the high selectivities that may be achieved in a wide range of chemical transformations. As mechanistic understanding of homogeneous catalysis has developed to an increasingly sophisticated level, a natural progression has been to extend successful applications in the manufacture of commodity chemicals and polymer intermediates to the production of fine chemicals. Economic and regulatory issues, together with environmental factors, have provided possibly greater stimuli for this change of direction. In the context of the environment, it is worth noting that the most polluting (per tonne of product) segment of the chemical industry is *not* the commodity chemicals but the pharmaceutical sector, which produces roughly 10^3 times as much waste as heavy chemicals, largely as a consequence of the use of stoichiometric reactions of poor selectivity.

The potential of enzyme-like selectivities inherent in homogeneous catalysts may be realized to the full in the synthesis of fine chemicals. Heterogeneous catalysts are conspicuous by their absence in this area of chemistry.

Global trends: commodity chemicals vs. fine chemicals

Since the early 1980s there has been a shift in the emphasis of many of the world's major chemical industries away from commodity chemicals and polymers towards anticipated high-growth markets in smaller-volume, high-added-value products. This approach has been seen as one means of maintaining profitability in the face of increased competition in commodity chemicals. A classic example of this trend was the sale by Monsanto of its highly successful methanol carbonylation business (Chapter **4**) in order to focus its assets on opportunities offered by life sciences markets.

Driven by demands for increased and sustained profitability there has been a shift in emphasis, by many of the world's chemical manufacturers, away from traditional commodity chemicals towards 'high added value' fine chemicals production.

Fine or 'speciality' chemicals include specialized polymers for electronics applications, intermediates for high-performance structural materials, and many biologically active compounds. The latter, which include pharmaceuticals, crop protection chemicals, flavours, fragrances, and food additives, have provided a particularly strong driving force for the use of homogeneous catalysts. This has been reinforced as a consequence of the eventual recognition, particularly by the pharmaceutical industry, that the application of organometallic chemistry and homogeneous catalysis can lead to economically viable alternative process routes that eliminate steps in the traditional and costly multi-stage procedures, based on conventional synthetic organic chemistry, that dominate drugs manufacture.

Economic factors: commodity vs. fine chemicals

Different economic factors dominate the production of fine chemicals relative to those prevailing in commodity chemicals manufacture.

Fine, effect, or speciality chemicals, unlike the traditional commodity products, have complex chemical structures and properties that justify a high selling price. In high added value situations such as these, where relatively small quantities of products are manufactured, factors such as catalyst costs, separation of product from catalyst and reactants, recycling/regeneration of catalysts, etc., assume reduced significance relative to that in the manufacture of commodity chemicals. This is simply because these costs can be more readily absorbed in the relatively high value of the products. Also, these are generally small-scale processes and the reaction steps tend to be carried out in a batch fashion rather than the continuous, and integrated, operation typical of the production of commodity chemicals and intermediates. The relative balance of economic factors is therefore rather different.

Asymmetric catalysis: regulatory factors

The ultimate application of high selectivities concerns the development of catalysts for the synthesis of chiral compounds with high degrees of enantioselectivity. The need for chiral specificity in bioactive products reflects the fact that most enzymes have inherent chirality ('handedness'). As a consequence, attempts to manipulate biological systems through therapeutic drugs or aroma enhancers often involve use of chemicals containing chiral centres. The desired biological activity is usually associated with only one of the two stereoisomers of a chiral compound. In the extreme case, exemplified by thalidomide, one optical isomer is therapeutic and the other has serious undesired biological consequences (teratogenicity). To improve product safety, the pharmaceutical industry is producing an increasing number of products in enantiomerically pure form and it is only a matter of time before legislation will demand this as a routine requirement.

Food and drug legislation to demand the production of single enantiomers as a routine requirement in developed countries is imminent.

Chemical advances

In the past many attempts to effect, for example, asymmetric catalytic hydrogenation led to products with disappointingly low enantiomeric excesses. This was undoubtedly a consequence of the complex nature of the surfaces of the *heterogeneous* catalysts, containing many types of reactive sites (*cf.* Chapter **1**), that were used initially. Developments in organometallic chemistry and homogeneous catalysis have been instrumental in improving this situation beyond recognition. The discovery in the mid-1960s that L-3,4-dihydroxyphenylalanine (L-dopa) was effective in the treatment of Parkinson's disease created a sudden demand for this rather rare amino acid. This coincided with the discovery of Wilkinson's catalyst, $RhCl(PPh_3)_3$, which opened up the possibility of selective hydrogenation of unsaturated hydrocarbons under mild reaction conditions. These developments, coupled with interest in the development of synthetic routes to organophosphines containing an increasingly wide range of substituents, enabled access to the possibility of catalytic asymmetric hydrogenation, simply by the substitution of PPh_3 with a chiral phosphine ligand. These factors led to the development of what are arguably the most elegant applications of homogeneous catalysis, namely the synthesis of optically active organic compounds from non-chiral starting materials.

More detailed mechanistic understanding, particularly of ligand effects in homogeneous catalysis, coupled with sophisticated ligand design, has facilitated the discovery and development of an increasing number of new homogeneous catalyst systems for the production of specific molecules in high enantiomeric excess.

Asymmetric induction can be achieved in many reactions catalysed by transition metal complexes but the first commercial application, introduced in 1974, was in fact the Monsanto synthesis of L-dopa. This represented not only a landmark in industrial asymmetric synthesis, but also provided a great incentive for research into additional applications of homogeneous catalysis for the synthesis of fine chemicals. Even so, relatively few homogeneous enantioselective catalysts have been commercialized to date. Representative examples of applications in fine chemicals manufacture are summarized in (Table 7.1)

Table 7.1 Some commercial applications of homogeneous catalysis in the manufacture of fine chemicals

Reaction	*Catalyst*	*Product*	*Use*
Hydroformylation	Rh	vitamin A	natural product
Hydrocarboxylation	Pd	ibuprofen	pharmaceutical
Hydrogenation of enamides	Rh/DIPAMP	L-dopa L-phenylalanine	pharmaceutical food additive
Isomerization of allylic amine	Rh/BINAP	L-menthol	aroma and flavour chemical
Epoxidation of allylic alcohol	$Ti(OPr^i)_4/Bu^tOOH$ di-*iso*-propyl tartrate	disparlure glycidol	insect attractant intermediate
Cyclopropanation	Cu/chiral Schiff base	cilastatin	pharmaceutical

DIPAMP = *bis*-{(phenyl, *o*-methoxyphenyl)phosphino}ethane
BINAP = 2,2′-*bis*(diphenylphosphino)-1,1′-binaphthyl

Homogeneously catalysed reactions often comprise *the* key enabling step in manufacturing routes to fine chemicals.

These are not exhaustive, but have been selected to illustrate both the range of catalytic reactions, e.g. hydroformylation, hydrogenation, isomerization, epoxidation and cyclopropanation, and their importance in key enabling reaction steps in the manufacture of specific products.

Chiral ligand availability

The development of synthetic routes to chiral, bidentate phosphine ligands having C_2 symmetry has been crucial to progress in asymmetric homogeneous catalysis.

The essential feature for selective synthesis of one optical isomer of a chiral substance is an asymmetric site that will bind a prochiral olefin preferentially in one conformation. The recognition of the preferred conformation can be accomplished by the use of a chiral ligand coordinated to the metal, the ligand creating what is effectively a chiral hole within the coordination sphere. An important factor in the successful application of homogeneous asymmetric catalysts has been the design and development of a range of chiral, usually *bidentate*, phosphine ligands, especially those having C_2 symmetry, for use with different metal centres. Some of the most successful examples are illustrated in **7.1**–**7.5**.

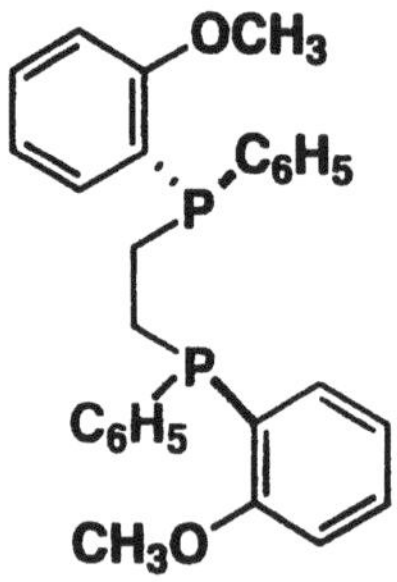

(R,R)-DIPAMP **7.1** (S,S)-CHIRAPHOS **7.2** (R)-BINAP **7.3** DUPHOS **7.4** (S,S)-DIOP **7.5**

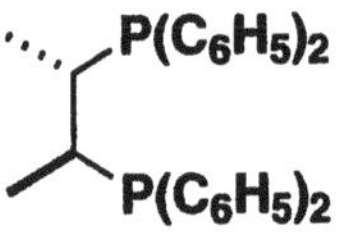

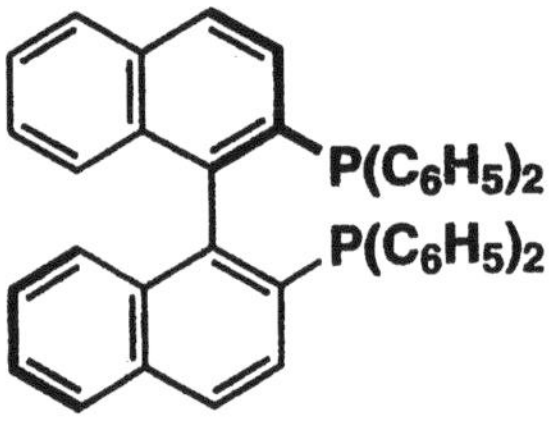

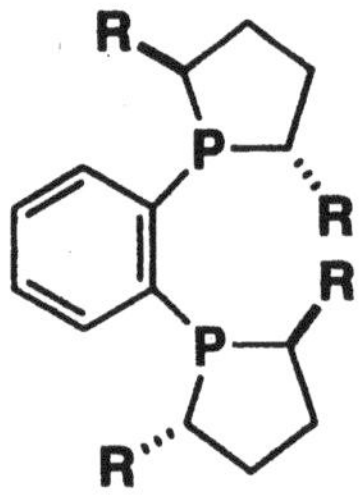

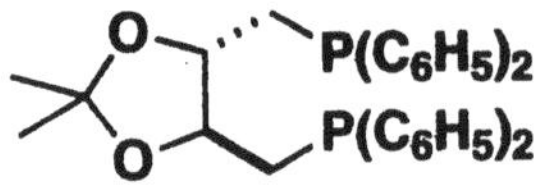

7.2 Vitamin A

The application of homogeneous catalysis in the synthesis of fine chemicals is of course not new. One of the earliest examples involves the use of hydroformylation as a key step in the synthesis of vitamin A, developed at BASF in the late 1950s; a 600 te y^{-1} plant has been operational since 1971. The synthesis involves a Wittig-type coupling between a vinyl-β-ionone (C_{15}) and γ-formylcrotyl acetate (C_5). The former is derived from linalool which itself is obtained from petrochemical feedstocks (*iso*-butene, formaldehyde, acetone and acetylene). The production of the C_5 fragment requires a key hydroformylation step (Fig. 7.1), in which 1-vinylethylenediacetate is converted into the corresponding *branched* aldehyde with regioselectivities of up to 80%, using a rhodium catalyst under high pressure conditions (600 bar CO/H_2, 80°C). Thermal elimination of acetic acid yields 4-acetoxy-2-methyl crotonaldehyde, which is then coupled with the C_{15}-ylide building block to form the C_{20}-vitamin A acetate. An alternative process developed by Hofmann–La Roche also includes a hydroformylation step and together these share most of the vitamin A world capacity of approximately 3000 te y^{-1}.

High regioselectivity in the production of the *branched* aldehyde (*cf.* Chapter **3**) in the hydroformylation step is key to the formation of the C_5 building block in the synthesis.

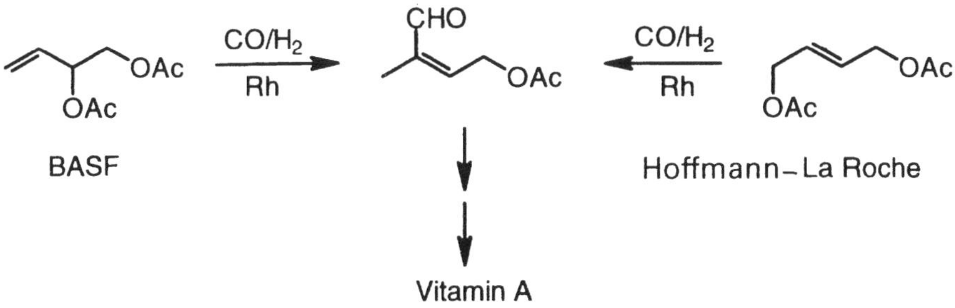

Fig. 7.1 The hydroformylation step in the synthesis of Vitamin A.

7.3 Ibuprofen

Two non-steroidal anti-inflammatory drugs ibuprofen (**7.6**) and naproxen (**7.7**) are members of the class of 2-arylpropionic acids towards which extensive synthesis research effort has been directed.

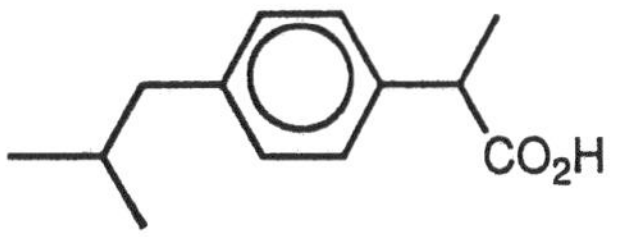

7.6 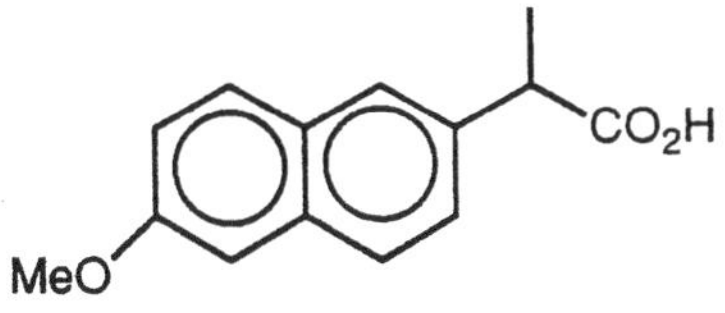 **7.7**

This reflects their multimillion pound per year market and the fact that previous synthetic methods were unsatisfactory. A Pd-catalysed hydrocarboxylation route to the former, using *p-iso*-butylstyrene, has been commercialized jointly by Boots and Hoechst Celanese. Furthermore, the use of a bidentate phosphine ligand, (R)-(–)- or (S)-(+)-binaphthyl-2,2-diyl hydrogen phosphate has been demonstrated to give optically active ibuprofen.

Another member of this family, (*S*)-naproxen, is one of the world's largest-selling prescription drugs. It is sold as the pure (S)-isomer because the (R)-isomer is a liver toxin. The desired isomer may be

This application represents one of the few potential applications of Ru in homogeneous catalysis (*cf.* Chapter **1**).

obtained by conventional optical resolution of the racemate. Many alternative routes have been explored but the most favoured one employs asymmetric hydrogenation. The synthesis starts with an electrochemical reduction of a methoxy-substituted acetylnaphthalene in the presence of CO_2. The resulting α-hydroxypropionic acid is dehydrated over an acid catalyst to produce the α-naphthylacrylic acid which provides the substrate for enantioselective hydrogenation using an (S)-BINAP Ru(II) chloride complex (Fig. 7.2). The reaction is carried out at 135 bar H_2 in the presence of excess triethylamine to give the required product in optical yields of 96–98%.

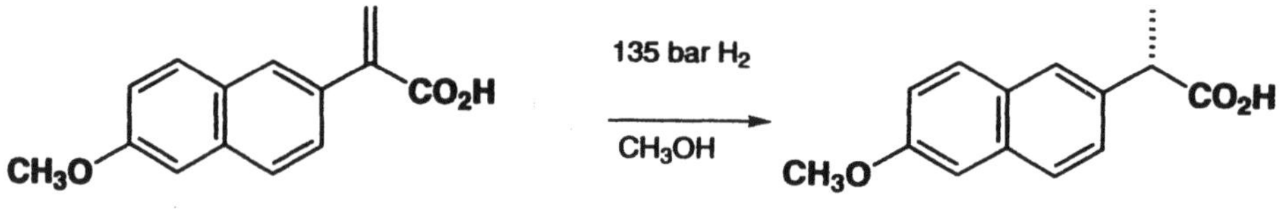

Fig. 7.2 A synthesis of (*S*)-Naproxen.

7.4 L-Dopa

The key step in the L-dopa synthesis is the catalytic hydrogenation of the prochiral olefin, a substituted acetamidocinnamic acid, the product of which is converted into the therapeutically effective isomer of dihydroxyphenylalanine in a subsequent step (Fig. 7.3).

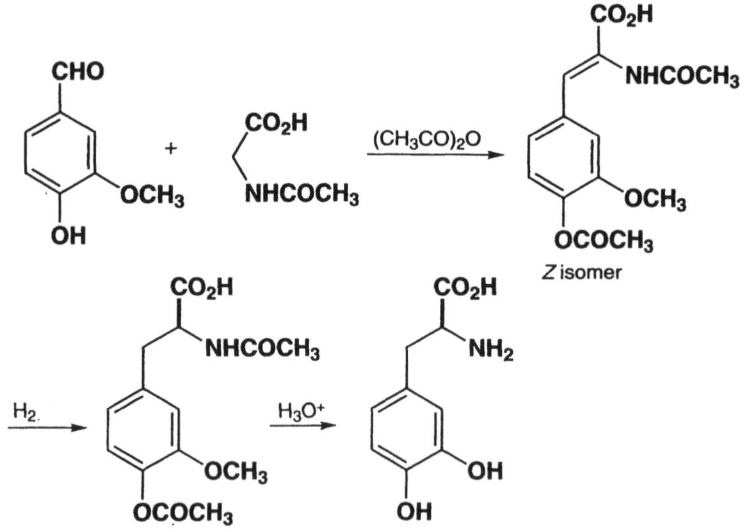

Fig. 7.3 The synthesis of L-dopa.

The catalyst is prepared in situ by the addition of the chiral chelating phosphine DIPAMP (**7.1**) to $[Rh(COD)_2]^+$ (COD = cyclo-1,5-octadiene) which is converted into a species of the type $[Rh(DIPAMP)H_2(ROH)_n]^+$ (R = Me, Et) under reaction conditions (3 bar H_2, 50°C, aqueous alcohol solvent).

Substrate to catalyst ratios as high as 20,000:1 can be tolerated. The hydrogenation is carried out in a solvent system in which both the starting material and product are insoluble but in which the racemate is soluble. Filtration after reaction gives the desired product in high chemical (*ca.* 90%) and enantiomeric purity (> 95% ee). The catalyst remains in the mother liquor from which the rhodium can be extracted and isolated for re-use. It is reported that one pound of catalyst yields one tonne of L-dopa. In systems of this type it is evident that the catalyst costs are high, not only because they contain Rh, but also because they contain chiral phosphines which are not only difficult and expensive to synthesize but are also difficult to recover and re-use. However, in high added value situations such as these, high catalyst costs can clearly be tolerated (Section **7.1**). This process required only intermittent operation to satisfy the world requirement for L-dopa and the technology was ultimately sold by Monsanto to May & Baker.

In the L-dopa process a *monodentate* chiral phosphine was used initially. However, subsequent research showed enhanced enantioselectivity with *bidentate* ligands and since then the latter have been preferred in essentially all applications of asymmetric homogeneous catalysis.

The product purity is significantly higher than that obtained by natural product synthesis and the catalytic process was so successful that it rapidly displaced more conventional routes based on synthetic organic chemistry.

A similar Rh-catalysed hydrogenation of an α-acetamidocinnamic acid has been utilized in the production of *N*-acetyl-L-phenylalanine in high optical yield. The demand for this product has increased as a result of the commercial success of the synthetic sweetener aspartame.

Reaction mechanism

An in-depth understanding of the enzyme-like specificity of enantioselective hydrogenation posed a formidable scientific challenge and led to intensive study, particularly of Rh(I) catalysts bearing chelating diphosphine ligands. Recourse has been taken to previous detailed kinetic and mechanistic investigations of the role of $RhCl(PPh_3)_3$ in the homogeneous catalytic hydrogenation of simple olefins (Chapter **2**). Nevertheless, some unexpected lessons have been learned from this exercise.

An in-depth mechanistic understanding of this reaction has been developed only after many years of effort.

It is believed that the acetamidocinnamic acid substrate coordinates to Rh through both its C=C bond and the C=O function of the acetamide grouping, a rigid arrangement which constrains the substrate to present one face preferentially to the metal atom. The transfer of hydrogen to the bound face of the olefin generates one optical isomer specifically. However, because oxidative addition of H_2 to the Rh(I) complex is usually rate-limiting, detectable quantities of the more stable olefinic complexes have been found to accumulate in solution. Perhaps the most surprising and significant conclusion from this mechanistic work, which provides an important take-home message to

The major product arises from the *less* stable initial olefin complex which is *undetectable* by spectroscopic methods, an important take home message to *all* practitioners of catalysis.

all practitioners of catalysis, is that the *major* product arises from the *less-stable* initial olefin complex which, in solution, is undetectable by spectroscopic methods. The rate-limiting reaction of the weaker complex with H_2 is clearly sufficiently faster than that of the stronger complex and dominates the kinetics of the overall process. It therefore determines the chirality of the product, since the oxidative addition of H_2 and subsequent steps appear to be irreversible.

7.5 L-Menthol

The largest-scale application of homogeneous asymmetric catalysis includes a key isomerization step in the synthetic route to L-menthol.

A very significant example in the development of asymmetric homogeneous catalysis concerns the isomerization of olefins. A synthetic route to enantiomerically pure menthol, one of the most ubiquitous fragrances and food additives (spearmint), which requires a key enantioselective isomerization step, has been developed by Takasago Int. Corp., Japan. It is operated on a 1500 te y^{-1} scale, and represents the world's largest application of asymmetric catalysis.

The isomerization reaction of an allylic amine to an enamine represents the first commercial application of the BINAP ligand, discovered and developed by Noyori.

The key reaction in the process appears to be the generation of the first chiral centre by this enantioselective isomerization step.

As observed with catalytic hydrogenation in the manufacture of L-dopa, a metal centre, to which one face of a prochiral olefin preferentially coordinates, should also be able to effect a stereoselective 1,3-hydrogen shift to produce specifically one of the two optical isomers of the product olefin. This possibility has been realized commercially for the isomerization of an allylic amine to an enamine. The key to success has been the use, once again, of Rh-based catalysts, in this case in combination with the BINAP ligand (**7.3**), one which has also proved useful in the enantioselective hydrogenation of olefins and ketones. The sequence of reactions involved in the production of L-menthol is summarized in Fig. 7.4. *β*-Pinene, a natural terpene, is first pyrolysed to form myrcene. Addition of Et_2NH catalysed by Bu^nLi yields diethylgeranylamine and its *Z*-isomer, diethylnerylamine. Either of these allylic amines may be isomerized to (*R*)-citronellal (*E*)-enamine, but the 'handedness' of the product depends on the chirality of the BINAP ligand present in the catalytic Rh precursor, [Rh(±)-(BINAP)(COD)]ClO_4. The reaction is operated on a 7 tonne batch basis at substrate:catalyst ratios of 8000–10,000:1. Complete conversion to the enamine (in 98% enantiomer excess), at turnover numbers (TON) of >400,000 occurs after 15 h at 100°C. The product is distilled directly from the reaction mixture at low pressure and the active catalytic residue can be re-used directly. In commercial practice the *p*-tolyl analogue of BINAP is preferred since it enhances catalyst stability and solubility, thus improving the degree of recycling. Hydrolysis of the enamine with cold aqueous acetic acid gives synthetic *R*(+)-citronellal, in an optical purity of 96–99%, which is much greater than the highest value of the natural compound (82%). As a consequence the complex multistep synthesis shown in Fig. 7.4 is

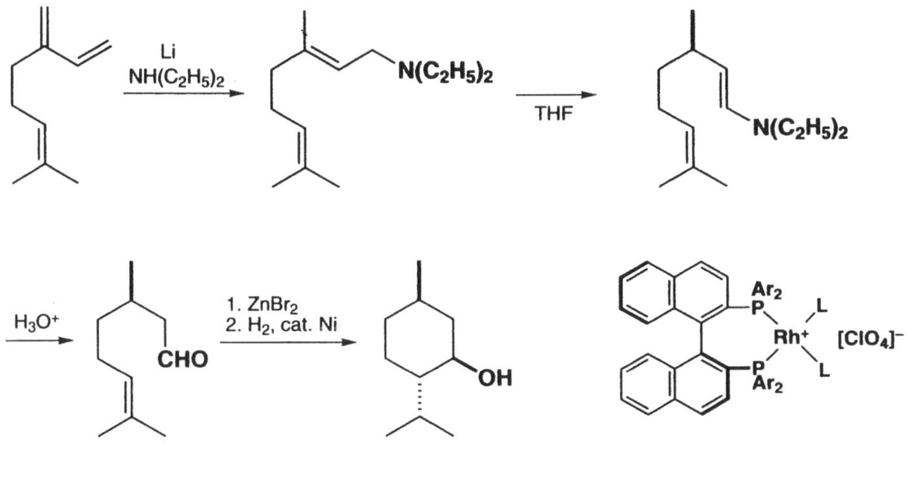

Li
$NH(C_2H_5)_2$
$N(C_2H_5)_2$
THF
$N(C_2H_5)_2$
H_3O^+
CHO
1. $ZnBr_2$
2. H_2, cat. Ni
OH
Ar_2
P
L
Rh^+
$[ClO_4]^-$
P
L
Ar_2

(*S*)-BINAP-Rh^+ catalyst; L = THF, acetone, 1,5-cyclooctadiene, (*S*)-BINAP

Fig. 7.4 The synthesis of L-menthol.

economically competitive with production of menthol from natural sources. The key appears to lie in the creation of the first chiral centre by the enantioselective isomerization reaction. The subsequent Lewis acid-catalysed ring closure is fortuitously stereospecific and gives a single diastereomer with the desired configuration. Reduction of the remaining alkene double bond over a conventional heterogeneous Raney Ni catalyst completes the menthol synthesis.

This menthol synthesis is all the more remarkable because three chiral centres are created, all of which are necessary to produce the characteristic menthol odour and local anaesthetic action.

7.6 Asymmetric epoxidation

Together with hydrogenation and isomerization, epoxidation completes the trio of commercially significant applications of enantioselective homogeneously catalysed reactions. Stereospecific olefin epoxidation is distinctive in that it creates two chiral centres simultaneously; it is also useful for kinetic resolution of racemic mixtures of chiral olefinic compounds.

Enantioselective epoxidation has the merit that itcreates two chiral centres simultaneously.

The discovery that Ti complexes containing asymmetric ligands can catalyse the enantioselective oxidation of allylic alcohols to the corresponding epoxyalcohols has been of considerable impact. Such catalysts are both highly selective and convenient to use, offer advantages over the chiral complexes of V and Mo discovered previously, and as a consequence are used commercially in the synthesis of several speciality products in which biological activity is confined to a particular isomer. The Sharpless oxidation employs a catalyst prepared by treating a Ti(IV) alkoxide with a tartrate ester such as the natural (+) isomer of diethyl tartrate. Although they are

Enantioselective Ti epoxidation catalysts offer selectivity advantages over V and Mo, but are limited to allylic alcohols as substrates.

rather sluggish catalysts (and are in fact used almost in reagent quantities in some transformations), and require the handling of organic peroxides on a large scale, they have the advantage of giving exceptionally clean, selective reactions. Restriction to the oxidation of allylic alcohols represents a major limitation of these systems; homoallylic alcohols are oxidized less cleanly and the oxidation of simple olefins shows little enantioselectivity.

Sharpless chemistry is utilized in a key asymmetric epoxidation step in the manufacture of (+)-disparlure, the gypsy moth pheromone.

Probably the best known of the applications of Sharpless chemistry is the synthesis of a chiral epoxide as an intermediate to (+)-disparlure, the pheromone for the gypsy moth, commercialized by J. T. Baker in 1981. The synthesis of (+)-disparlure employs the enantioselective epoxidation of Z-2-tridecan-1-ol with *t*-butyl hydroperoxide at −40°C for 4 days in the presence of a complex derived from $Ti(OPr^i)_4$ and (S,S)-diethyl tartrate as catalyst. The resulting epoxyalcohol is formed in 80% yield and 90–95% enantiomeric purity before recrystallization. Further conversion to (+)-disparlure requires three subsequent conventional reaction steps via an intermediate aldehyde. The introduction of this asymmetric epoxidation route on the multi-kilogram scale reduced the price of disparlure by an order of magnitude. The very high activity of disparlure suggests that a production capacity of only a few kg y^{-1} is required to satisfy demand.

(+)-disparlure is used by the US government for insect control and it is notable that the federal revenues saved in just two years exceeded that expended on research grant funding the 10 year search for the method!

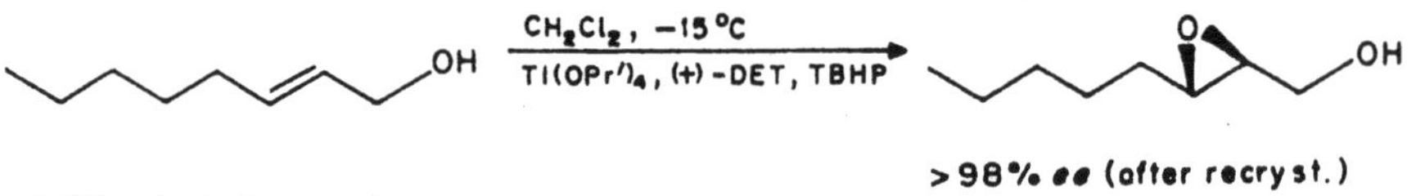

OH $\xrightarrow[Ti(OPr^i)_4,\ (+)\text{-DET, TBHP}]{CH_2Cl_2,\ -15\,^\circ C}$ OH

>98% ee (after recryst.)

DET = diethyl tartrate

TBHP = *tert* - butyl hydroperoxide

Fig. 7.5 Synthesis of C_8 epoxyalcohols using asymmetric allylic alcohol epoxidation.

A significant breakthrough, and the basis for a truly economical use of enantioselective epoxidation on a larger scale, came with the recognition (1984) that the addition of molecular sieves to the reaction mixture allows the use of 5 mol% Ti/tartrate complex in a truly catalytic and highly reproducible manner. Upjohn (1985) and ARCO (1988) have both utilized this development as the basis of commercial routes for the production of C_8-epoxyalcohols in > 98% (Fig. 7.5) and for up to 10 te y^{-1} of (*S*)- and (*R*)-glycidol, in 88 ± 2% enantiomer excesses respectively. The latter molecules are used as basic building blocks for many highly functionalized chiral molecules; (*S*)- and (*R*)-methylglycidol are also produced by this route.

A derivative of Sharpless chemistry is also used commercially for the manufacture of C_8-epoxyalcohols and both enantiomers of glycidol.

A more recent alternative approach, developed by Jacobsen and co-workers, concerns the catalytic asymmetric epoxidation of *un*functionalized olefins using (cheap) NaOCl as oxidant in the presence of Mn complexes of chiral Schiff bases (**7.8**) as catalysts.

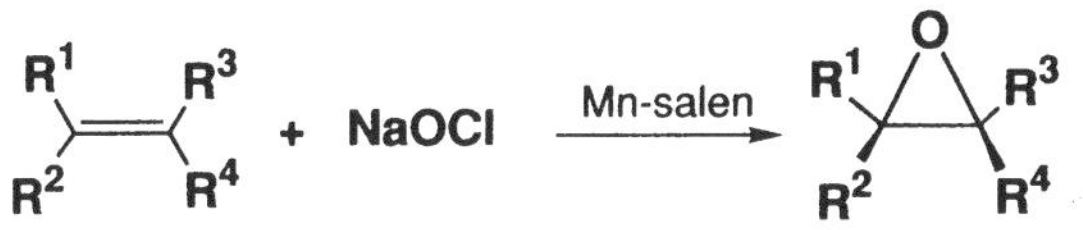

$$R^1R^2C{=}CR^3R^4 + NaOCl \xrightarrow{\text{Mn-salen}} \text{epoxide}$$

catalyst =

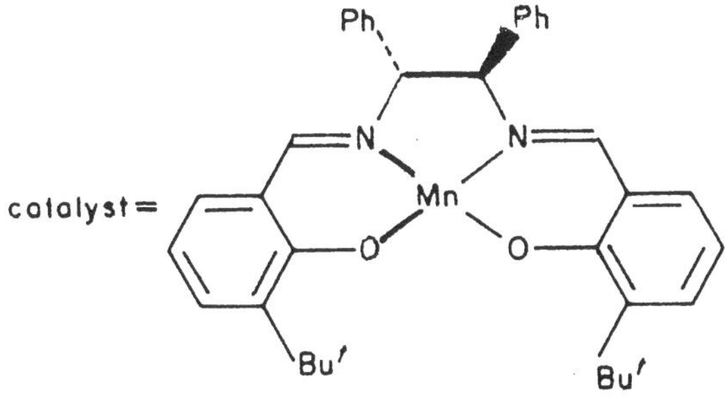

7.8

Optical yields of up to >97%, depending upon the nature and positions of the substituents, have been reported. The asymmetric epoxidation of some *cis*-olefins which lead to enantiomerically pure aminoalcohols for the synthesis of K^+ channel activators, HIV protease inhibitors, and anticancer drugs has reportedly been scaled up to operate in a 6000 litre reactor. Potential problems to full-scale commercialization include the synthesis and availability of the olefins, the availability of the Schiff base (salen) ligands on a large scale, and the activity and stability of the catalyst.

In principle, the Jacobsen route provides greater flexibility over Sharpless chemistry by offering the possibility of catalytic asymmetric epoxidation of unfunctionalized olefins.

7.7 Cilastatin

One of the earliest asymmetric syntheses to be shown to proceed in very high enantiomeric excess is cyclopropanation, from the reaction of olefins with alkyl diazoacetates in the presence of Cu complexes containing chiral ligands. Even here the lead time to industrial exploitation was almost 20 years. A key step in the Sumitomo process for the production of cilastatin (Fig. 7.6), incorporates such chemistry. Cilastatin is used in conjunction with imipenem (a broad-spectrum antibiotic) to suppress hydrolysis of the latter by renal enzymes. An efficient reversible inhibitor of this enzyme is cilastatin, and a combination of imipenem and cilastatin now provides a very useful pharmaceutical.

Asymmetric cyclopropanation represents one of the earliest examples of highly enantioselective transformations.

The precursor to cilastatin, ethyl (*S*)-2,2-dimethyl cyclopropane carboxylate is prepared, in 92% enantiomeric excess, by the decomposition of ethyl diazoacetate with *iso*-butene using a dimeric

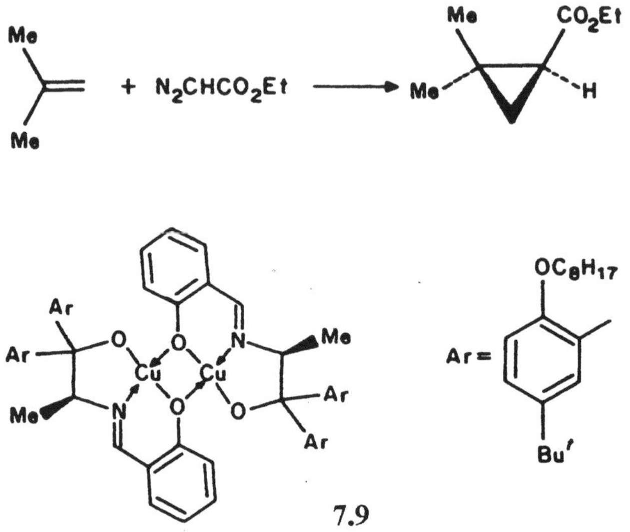

Fig. 7.6 Asymmetric cyclopropanation in (+)-ethyl (*S*)-2,2-dimethylcyclopropanecarboxylate synthesis, a key intermediate in the production of cilastatin.

chiral Cu catalyst (**7.9**) containing a Schiff base ligand. The distinctive feature of this reaction is that the catalyst creates a carbenoid fragment from the diazoester and then adds it selectively to one face of the olefin. Several types of catalyst are effective for this sort of cyclopropane formation, but the enantioselective character of the Cu-catalysed reaction is remarkable. Clearly, access to many other chiral cyclopropane derivatives is available using this technology, particularly attractive candidates being analogues of the naturally occurring pyrethroid insecticides. As yet, however, no further commercialization has been reported.

The search is still open for economically viable asymmetric cyclopropanation routes to analogues of naturally occurring pyrethroid insecticides.

7.8 Future prospects

It seems likely that the principal reactions discussed in the preceding section, namely asymmetric hydrogenation, isomerization, and epoxidation, will ultimately find extensive use in the production of pharmaceuticals, given the regulatory trend towards the treatment of enantiomers of the same compound as distinct therapeutic agents. This may also create commercial applications for the enantioselective osmium catalysed vicinal hydroxylation reaction. Clearly the complex chemistry in this area comprises a relatively young discipline, but there can be no doubt that commercial applications of enantioselective homogeneous catalysis are set to increase rapidly.

The potential for commercial applications of asymmetric homogeneous catalysis is enormous, and it is fitting to note that the achievements of pioneers in this area, Knowles (Monsanto, L-dopa), Noyori and Sharpless, have been recognized by the award of the 2001 Nobel Prize for Chemistry.

Further reading

A number of excellent texts provide more comprehensive coverage of much of the material described here.

Applied homogeneous catalysis with organometallic compounds, B. Cornils and W. A. Herrmann, eds., VCH, Weinheim, Germany, 1996, Vols. 1 and 2.

Industrial organic chemistry, K. Weissermel and H. J. Arpe, VCH, Weinheim, Germany, 3rd Ed., 1997.

Industrial organic chemicals, H. A. Witcoff and B. G. Rueben, John Wiley & Sons, New York, 1996.

Transition metals for organic synthesis, M. Beller and C. Bolm, eds., Wiley-VCH, Weinheim, Germany, 1998, Vols.1 and 2.

Organometallics 1 and 2, M. Bochmann, Oxford Chemistry Primers, Oxford, 1994.

Homogeneous catalysis, G. W. Parshall and S. D. Ittel, John Wiley & Sons, New York, 2nd Ed., 1992.

Catalytic chemistry, B. C. Gates, John Wiley & Sons, New York, 1992.

Principles and applications of organotransition metal chemistry, J. P. Collman, L. S. Hegedus and R. G. Finke, University Science Books, Mill Valley, California, 1987.

Industrial applications of organometallic chemistry and catalysis, J. Chem. Ed., 1986, **63**, 188-225.

Index